FROM SWORDS TO PLOWSHARES

The Path to Global Peace

BY DENNIS HAUGHTON, M.D.

Library of Congress Cataloging-in-Publication Data

Haughton, Dennis (Dennis L.)
From Swords to Plowshares.

Bibliography: p. 185
1. Peace. 2. Nuclear disarmament.. 3. Chernobyl Nuclear Accident, Chernobyl, Ukraine, 1986. I. Title
JX1953.H38 327.172 87-31107
ISBN 0-933703-96-1

Printed in the United States of America

First Edition

TABLE OF CONTENTS

PREFACE

A short while ago I sat on my rooftop garden under the immense dome of the heavens and watched the nightly spectacle of light in the western sky change from orange to purpleblue as the sun slipped away to begin someone else's day. As the sliver of the new moon and then Jupiter and its moons appeared, I became aware of the greater system of worlds that we are a part of. As the purple deepened to black and the stars emerged, I could also see Mars, Saturn and with a little imagination, for I knew where they were hiding, Uranus, Neptune, and Pluto. I could feel the rotation of the Earth beneath me and sensed the motion of the planets as we circled the star from which we were born. Then for a brief moment, I felt and saw the delicate web of energy that connected these worlds to me and every part of our galaxy and the universe beyond. Pulsating and alive, the Life Force flowed through me and reached out and touched all the distant suns and worlds in the unfathomable reaches of infinity.

We often forget that the Life Force joins us to each other in the same way. We all belong to the same family of humankind on this planet, although many have not yet awakened to that awareness. Sometimes we get so involved with our separate lives, our differences, and our limited realities that we forget our greater connection to the whole. Whether you are a fisherman in Seattle, an accountant in Boston, a billionaire in Buenos Aires, the president of a great nation, or an illiterate Bushman in Africa who can't even read this, we all have our parts to play in the same life flow. We share the same world, breathe the same air, and when we look up at night at the firmament above, we will begin to see that our destinies are intertwined. We are in the process of coming into a new awareness: a growing consciousness of one people, joined.

Somewhere out there, other intelligent beings must gaze out at the same heavens and contemplate their connection to the life flow also. It is likely that 5000 worlds just like this one flourish with life in our galaxy alone. No matter where or on what world life evolves, there must come a time when the intelligent species becomes capable of leaving its home world and reaching out to become part of the greater web of life spanning the universe.

Our race is now on the threshold of that reality. We are poised on the dawn of a new age for mankind. Never before in our history have we been at this juncture. In the next few decades our children will be developing the knowledge that will allow our grandchildren to live beyond the familiar confines of Earth: perhaps first in an orbiting space colony or permanent lunar settlement, and in due time they will reach out toward the stars themselves. It is important to me that we leave our heirs a world to which they are glad to return.

The advent of space travel and instant global communication has given us a new and expanded consciousness of who we are. Our astronauts have told us that once you see our planet from space in its wholeness, you no longer think of yourself as simply an American or Russian, but as a part of Humanity. We are on the verge of a planetary shift in consciousness that very rapidly will bring about a phase transition into an entirely new age of mankind—one that has been foretold by visionaries and prophets throughout history.

As intelligent races go, however, we are but in our infancy. Just barely into the age of technology, we have unlocked secrets of the universe before we possess the wisdom to use them safely. The great visionary Einstein knew when he gave us $E=mc^2$ that it would be a great challenge for us to use it intelligently. The secret of the atom perhaps is one of the crucial trials a race must pass on its way to wisdom. Races that don't learn to live peacefully with each other or who put power ahead of survival, annihilate themselves before they can muck up the rest of the universe. We now stand at that crossroad. The choice is ours—live peacefully with one another or perish.

What are the challenges we will need to solve together? Most of them are not new. Priority number one is ending war. Throughout recorded history, we have fought and killed one another in the name of God, or power, or simply because one group felt superior to another. Our weapons of destruction are now so powerful that their use quite possibly could end our civilization and other life forms on the planet. Locked in a deadly and pointless power struggle, the superpowers have amassed enough atom bombs to destroy a million Hiroshimas. Roughly equal to several thousand times the power of all the bombs used in WW II, we can now fry each other to a crisp many times over and yet some warlords talk of needing more.

The social costs of supporting our global military habit are staggering. Between 1960 and 1985 the world spent 14 trillion dollars on war related activities. Today we throw away about a trillion dollars a year—money that is unavailable for peaceful uses. Although we produce enough food, by some estimates 700,000,000 of our people don't get enough to eat. Malnutrition and inadequate health care kills 15,000,000 infants and small children a year, 41,000 a day. None of this is necessary, yet we permit it, for we have given armaments a higher priority than life. For roughly what we spend in 1 minute on the arms race, we could save 4 million children a year with immunizations that cost only fifty cents per child.

We live in a world of dualities: good/bad, black/white, right/wrong, win/lose, East/West, democracy/communism, them/ us, life/ death, and so on and so on. Neither wanting to be in second place, the superpowers compete for control over developing nations. One side fosters the rights of the individual, the other promotes the ideals of the collective. Both have their strengths, both their weaknessess. Both permit the abuse of power by individuals who have gained special privileges by way of political office, money, prestige or ruthlessness. We still are stuck in the mentality of duality that says if one side is right, the other must be wrong. If one wins, the other loses. Seen in this light, the world will be forever locked into deadly conflict until one system has eliminated the other. There is, however, a deeper reality beyond duality in which we can all benefit, we can all win. The path to this new understanding, as Patricia Sun teaches in her workshops, is easy—it requires only a mental shift in awareness.

In our shortsightedness and ignorance we have been careless and irresponsible with our planet's life support system in recent years. With seeming disregard for the consequences to those who come after us, we have been depleting our forests, poisoning our waterways, befouling our atmosphere, making large areas of our land uninhabitable with our excrement, and in general threatening to make a mess of things that were much more in harmony with nature before we came along.

There is a balance in life that we have not yet mastered. Aside from the nuclear issue, byproducts of our advancing technology threaten to bring about harsh changes in our environment. As we burn up our fossil fuels and increase the levels of atmospheric carbon dioxide, we alter the heating effect of solar radiation. Some say this will cause a greenhouse effect, leading to warmer global temperatures, melting polar ice caps, and rising oceans that would threaten our coastal cities. Others say we will precipitate lower temperatures and trigger another ice age. Either scenario could drastically change weather patterns and hinder our ability to grow enough food. Malignant skin cancers may affect one out of three adults by the early 21st century because we are depleting a thin layer of ozone in our atmosphere that filters out harmful ultraviolet rays. We are allowing lethal radioactive poisons to escape into our environment that can be harmful to life for thousands of years to come.

To solve the challenges that face us as a people, we will need all of our resources and peaceful cooperation among our nations. Squandering vast amounts of our limited assets on war machines destined to sit idle—because their use would be suicidal—is illogical to say the least. Thus, our first priority is turning our swords into plowshares and learning the art of global cooperation. Humanity can no longer pursue dangerous and narrow-minded nationalistic goals at the expense of planetary needs as a whole.

Now I don't know about you, but I for one have grown tired of the same old excuses for not having peace that politicians are forever bemoaning about. It seems it's always the other guy's fault. For the entire 40 years of my sojourn here, the U.S. and the Soviet Union have been at war with each other—perhaps not in the usual sense of shooting bullets and dropping bombs—but nonetheless, definitely not at peace. Certainly, the existence of tens of thousands of nuclear warheads pointed at each other is a clear sign of a disturbed relationship. The fact that both societies seem willing to bankrupt themselves in a pointless competition that could end up getting everyone killed seems a strong indication of underlying psychopathology on a massive scale.

Understanding this central example of racial insanity seemed crucial to my understanding the more general lack of peace in the world at large. So I asked myself two simple questions: why does this insanity exist, and can I do anything to help bring it to an end? The negotiating teams from both sides continue to struggle for a solution but they seem to keep getting bogged down in trivial details. Besides, we seem to be approaching the process backwards: trying to dismantle the war machine without first deciding to live peacefully with one another. It is not the nuclear weapons that are the problem. It is the underlying fear, mistrust, lack of genuine communication, and internal distresses that created these weapons. We won't create lasting peace if we deal only with the symptoms and not the cause.

When I realized that I really had no control over anyone else's thoughts and reality except my own, I began looking inward for the peace that seemed so elusive in the world outside. That inner search has spanned the good part of twenty years. The knowledge gained during that journey within has shown me that indeed peace is possible. Far from being off in hyperspace somewhere, the kingdom of heaven is found within. For me to glimpse that peace inside, I first had to acknowledge, accept, and forgive the raging war I carried on inside myself. I alone was responsible for the lack of peace in my being. You see, I found that peace as well as war are created first inside, and only then are they manifested outside.

This book is the story of my own personal voyage in seeking a peaceful relationship with the Soviet people. It was important that I get to know face to face the people that were supposed to be my avowed enemies. Because of the experience I shared with the Soviets, a lot has happened to me that perhaps wouldn't have otherwise. In the process I learned some very remarkable things, as well as getting caught in the fallout from the Chernobyl accident. I never planned on writing a book when I took off with the peace delegation in 1986 from L.A. International airport, but I didn't expect to come home radioactive either. The personal growth triggered in me by Chernobyl's radiation far outweighs whatever harm may come to me from it later on.

One night last year I sat down and wrote a letter to President Reagan and General Secretary Gorbachev, giving my heartfelt support to their efforts in continuing their dialog. When I realized the enormous odds against my letter actually reaching them personally, I began thinking of alternative ways of influencing them. It was then that I conceived the idea of writing this book. Not that the copies I have already sent to them will bring about drastic breakthroughs in international relations, but maybe if enough people wake up to the vision of possibilities created by a peaceful planet, then perhaps our leaders also would see beyond the shortsightedness of just reducing missiles. Maybe other people in their own way would be stimulated to spread the vision of a world without war and thus contribute to the healing of our planet.

I am not a trained diplomat. I, like you, am just a citizen. I have joined a growing number of ordinary people from all walks of life who are committed to bringing about global peace. Citizen diplomacy is not new, nor is it limited to any one country. In the United States alone there are an estimated 9,000 grass roots peace groups. We work independently of each other, although coalitions and national and international cooperation are becoming more frequent. We subscribe to no specific doctrines but share the belief that our present dependence on powerful military technology to guarantee "peace" is woefully inadequate. We are growing in awareness of ourselves as citizens of one world whose needs and desire for global harmony transcend all national boundaries, religious dogmas, or political creeds.

What I have learned in my search for peace, and especially in my experience since Chernobyl, is that grass roots efforts *can* make a difference. We don't have to wait for official diplomacy or our political leaders to bring about a safe planet. Only in believing in our own powerlessness will we continue to be ineffective and fail. You too, are important. You don't even have to leave your hometown to bring about change, for the roots of global disharmony are everywhere. All you need to do is to desire peace in your heart and then take whatever action brings you closer to it. Humanity, in its thousands of years of living together, has accumulated all the knowledge and wisdom necessary to bring about lasting peace. We only need to take the first step which is to know and believe that we can do it.

FOREWORD

There is a way to create peace that does not involve violence, coercion, or lies.

One day I was meditating very deeply, very earnestly, wanting to know, "What am I supposed to do in the world?" Then vividly I saw in my mind's eye, "TO END WARS." "To end wars," I thought, "Give me a break—this is too much." Then I thought, "Well, Patricia, don't get so carried away, find out what that means." As I returned to that meditative place and asked, "What does that mean?" I saw an incredibly beautiful vision of the planet, like you might see from outer space. And it was so beautiful and so radiant and so vibrant that my heart ached. I looked at it and I could feel its aliveness and I felt tremendous love for it. Then the words rolled up, "THE AGE OF AVATARS." An avatar is someone who realizes, feels and lives God, like Jesus or Buddha. And as I looked I saw gold-white light in cylinders coming down to the earth. Then, as I looked at one of these lights, I saw a beautiful face which I couldn't identify as male or female but simply as very beautiful and very radiant. Looking into the face, I realized and felt that this person had learned how to love themselves completely while in a body. Their compassion and understanding was complete. Their empathy was so authentic, so profound, that it radiated from them. When they looked at someone who was not "lit" they felt and saw each other, and the person who was not lit felt the person's love, realized it themselves. Then they became a light, and turning to someone else who wasn't lit, looked at them with such love and understanding that they felt it and then they became a light and it was happening all over the planet in little different places. It wasn't the whole planet, just parts, but then there was a critical mass reached, causing a quantum leap of love. Then the whole planet was lit and the words rolled up, "20 YEARS."

This vision, which I had about 15 years ago, began the work which I am doing now. I realized, for one thing, that the words "20 YEARS" could be interpreted various ways that suggested an evolutionary leap. Moreover, I realized from my experience of that vision that we didn't have to convince anybody of anything, we didn't need a new political party, we didn't need a new religion. In fact, we need to transcend the mental limitations which have us believing we need a "new program." And in order to transcend that, we only had to love and forgive ourselves. We only had to heal the fear in us that distances us from love and life. And just working on that would be the most powerful thing one could do to transform the world. The truth is, that is all you really can do anyway. Because the rest of what we do is an illusion, a "projection," pointing to others to change them.

Real transformation comes when you transform the only piece of the planet you have real power over, and that is the piece you are sitting in. I assure you, there is sufficient material there to work on.

We must dare to be, to grow up, to have compassion, to cultivate wisdom, to heal the planet, to create liberty from within ourselves. Let being authentic and wholesome be your guide. If you face and forgive yourself, you will become whole... and that can change the world.

Patricia Sun

CHAPTER ONE

An Introduction

On April 26, 1986, I was with a group of 60 Americans who took off from LA International and Kennedy Airports, on a journey to the Soviet Union that would turn out to be far more than we had bargained for. Little did we know that we were going to be caught in the middle of the worst nuclear disaster that humanity has yet created. Now as I look back on it all, I am happy that I went at that time. Not only has the experience caused profound changes in my own outlook on the world, but I think the world as a whole has grown wiser because of the accident. Chernobyl was only a warning to us. Clothed in the tragedy was a gift to our race—out of the ashes has arisen a renewed public awareness of the likely consequences of using nuclear military might to settle our differences.

The world situation at the time was filled with conflicts between nations as usual. At the same time, U.S. - Soviet relations were undergoing a long-needed warming trend. President Reagan and General Secretary Gorbachev had met in Geneva the previous December and optimism for two additional summits was still running high. Indeed, there seemed real hope of achieving significant arms reductions in a world whose armories were already overflowing with enough weapons to kill every person and creature on the planet many times over.

Dennis L. Haughton, M.D.

Several weeks prior to our trip, the U.S. had dismayed much of the world by bombing Libya. Regardless of the causitive events that preceded this action or whether or not it was justified, it was nonetheless perceived by many as a hostile act. World tension level rose in general, and U.S. - Soviet relations in particular were strained. In Moscow when we heard of a Soviet carrier headed for Libya, my own anxiety level soared as I imagined the possible consequences of the superpowers on a collision course.

Then suddenly there was Chernobyl, and Libya was soon forgotten in the ensuing deluge of publicity the frenzied media churned out about the reactor incident. One moment we were struggling to repress our nuclear anxieties over the Libya conflict, and the next we had a very real taste of nuclear holocaust to remind us of just what we were fooling around with.

Over a year has come and gone since then and, except for those of us who were directly affected, business goes on as usual. I know I will never be the same. I have done much soul-searching and thought a lot about our possible destinies since then. Our alternatives as a race are not all bleak. Only if we continue to live by the present social consciousness based on a world divided against itself are we destined to continue repeating the same old destructive patterns. We can, however, choose a much more desirable future by simply changing our consciousness and thinking processes. By changing our thoughts within we can change our belief in where we are going and thus manifest a much grander destiny.

There once were six blind men who met an elephant for the first time. One of them, holding its tail, declared that an elephant was like a rope. Another, feeling one of its legs, said that an elephant was like a great tree trunk. The one holding its trunk declared that an elephant was like a huge snake and so on. Like this, we have been perceiving reality for eons with only limited vision. We have created hundreds of religions to explain our relationship to a deeper spiritual reality of our universe. The predominant political ideologies of our time, democracy and communism, simply reflect different views of bringing order to social institutions. Like the blind men's view of the elephant, each of our belief systems is only a partial truth of something much larger.

What appears as a terrifying snake in the dark is quickly seen as a harmless rope on the ground in the illumination from the light of a candle. In the same way, the solution for a seemingly insolvable problem, such as how to achieve peace between two systems as radically different as the U.S. and the Soviet Union, becomes obvious when looked at from a more enlightened perspective. I like the vision of possibilities I have seen that has grown out of my new understanding of the Soviets as people.

I think that the group I went with made a big difference. My trip was sponsored by The Institute of Communication for Understanding (ICU) and as such we traveled as peace delegates. Unlike the people on a lot of "tours," many of us were there not only to see the usual sights, but also to meet and talk to ordinary Soviet people. We had scheduled meetings with peace committees, women's groups, and the like but, more importantly, many of us ventured off by ourselves, using the Soviet's excellent public transportation, and were able to meet many Soviet citizens. Contrary to preconceived expectations, we were not restricted from traveling freely in the cities we visited, nor, to our knowledge were we followed by KGB agents, as we had speculated we might be.

Last but not least, we were blessed to have Patricia Sun as group leader and spiritual guide throughout our sojourn. I still can't find the right words to describe Patricia. I have known her for more than ten years. She has catalyzed many realizations and much spiritual growth in me, and has helped me understand unconditional love and selfacceptance. Through her, I have learned a dimension of healing more fundamental than the knowledge gained in four years of medical school. Traditional western medicine focuses on killing diseases with drugs or surgery, Patricia teaches people to unlock their own inner healing energy through unconditional love. By integrating our logical linear left brain functions with our intuitive, unbounded, psychic right brain, we are able to transcend the illusion of duality and become whole. Like many natural healers, Patricia puts her hands on a person and by channeling healing energy, the people she touches get better.

Once, in Leningrad, I sprained my ankle running for a bus one morning. I knew by the amount of bruising and swelling that it was a bad injury. In fact, as a doctor, it was the worst looking ankle I had seen in over a year. Thinking that I may have needed crutches for the rest of the trip, I asked Patricia to "do her thing" on my ankle. Feeling nothing unusual as she placed her hands around it, I found myself becoming skeptical. The next day however, although my ankle was still massively swollen and bruised, I had no pain and could walk normally. That day I walked all over Leningrad and even went out dancing that evening without any problems.

Like the blind men and the elephant, our understanding of reality is incomplete when viewed from a limited perspective.

As our awareness expands, our previous "truth" is seen to be only one of several perceptions of reality.

Finally, as our growing awareness allows us to see the complete picture, we are able to understand how each limited reality is an integral part of the whole. As the evolutionary leap now in progress gives us a more expanded awareness and we begin thinking globally, radically different ideologies such as democracy and communism, Christianity and Buddism, will be seen as different expressions of the greater reality of humanity.

In a way, Patricia helped inspire me to write this book. Ever since I heard her talk many years ago about her inner vision of the world awakening (described by her in the forward of this book), I have witnessed bits and pieces of this vision becoming reality. Shortly after arriving in Kiev, a group of us were sharing our mutual excitement over our encounters with the Soviet people so far. Over and over again we were impressed with their warmth and genuine desire for peace. Rather than being at odds with each other we found a growing spiritual closeness arising from our shared desire for friendship. In a rather intense encounter with Patricia, while relating her vision to what we saw happening around us, I suddenly felt an electric wave of elation spread up my spine and cause my hair to tingle. Before our very eyes we were witnessing the realization of that vision. I looked into her eyes and said, "It's happening, isn't it? It's really coming about," and I knew that she saw it too.

For many years world peace has been an elusive and distant dream for much of humanity. Even as this is being written, we are killing each other in over 20 different wars taking place on various parts of our planet. But, as we rapidly approach the end of the twentieth century, a fundamental shift in collective world consciousness that has already begun will accelerate and bring us closer to fulfilling our dreams for peace. New and creative solutions to our mutual global problems will arise as expanded awareness allows us to see things from a more enlightened perspective.

There are prophets of doom around us who still await a coming Holocaust. In the years ahead not all transitions will be smooth. We may still experience global crises. We may yet see new hostilities break out or the sudden flaring of old ones. Global thermo-nuclear war, however, will never happen. It is possible that nuclear weapons could be used on a limited basis in the Middle East or elsewhere if deepseated conflicts escalate out of control. Perhaps we will need one last vivid reminder of the devastating consequences of nuclear fission unleashed against each other before we put away our weapons for good. And as for Armageddon, we have already lived through it: World War I, World War II, the concentration camps of Dachau and Auschwitz, Nagasaki, Hiroshima, Chernobyl, and so on. Whatever atrocities we haven't already committed against each other we have imagined in our nightmares or acted out in our movies. We need no further reminders of our potential for inhumanity. In experiencing the Holocaust in our collective consciousness, we have gained the wisdom to move on.

In the chapters that follow, I will explore the reasons for today's conflicts between Soviets and Americans, the world's leading adversaries. What has occurred between them applies equally to enemies the world over, for the basic process is the same. We will look at the opportunities for peace between the superpowers based on a new understanding of who we are. And then finally, as we sift through the ashes of Chernobyl, we will contemplate a new vision of our destiny together as our planet heals and becomes whole.

CHAPTER TWO

Overcoming prejudice and breaking the propaganda barrier

If peace is to become a reality on our planet, we must first shed the misconceptions and prejudices of each other that prevent understanding and harmonious coexistence. Just what do Soviets and Americans really know about each other? Do the prevailing social attitudes we have of each other reflect unbiased rational conclusions or are they more based on decades of dogmatic political rhetoric? Prior to visiting the Soviet Union I assumed to some degree that unbiased information about the United States would be hard to find in a closed society with a stateowned media. I also thought incorrectly, that in the United States, a country with a free press, propaganda would be more limited. Now after having been to the other side of the Iron Curtain and back, I conclude that disinformation (as propaganda is now termed in Washington) is pervasive on both sides and plays a major role in perpetuating the present arms race.

For the last 40 years, since the end of World War II, the two governments have waged a massive propaganda war with each other. Each has portrayed the other as a fundamentally inferior, flawed, and malevolent system that breeds the worst possible types of incorrigible scoundrels, thugs, hooligans, and other unwanted specimens of human behavior gone bad. Each has been quick to criticize and point out the other's obvious shortcomings and glorify its own achievements by comparison. This name-calling has been going on for such a long time that to a large extent it is accepted as "truth" by both societies. In short, we all have been brainwashed.

From early childhood, U.S. and Soviet citizens are bombarded with a multitude of written, verbal, and visual images of each other that too often reinforce our misleading and stereotypical images we have of one another. Although Soviet school children are exposed to a wide range of American authors such as Edgar Allen Poe, Washington Irving, Jack London, John Steinbeck, Walt Whitman, Mark Twain, Eugene O'Neill, and Ernest Hemingway, their history books often dwell on the negative aspects of American culture. American school books largely ignore Soviet history and, with the exception of Tolstoy and Dostoevsky, American children grow up ignorant of most Soviet literature.

Sometimes the propaganda has been blatantly obvious, but in the fervor of patriotic pride it has often been accepted as unquestioned truth. As part of the research for this book, I had the opportunity to watch some of the old newsreels made in the fifties when the American anti-communist campaign was more crude. "Duck and cover" drills were taught to school children in case the atom bomb was dropped by the "Reds" or "Commies", as the Soviets were usually labeled then. As sirens shrieked, boys and girls dove under their desks and put their little hands over their heads to ward off the oncoming thermonuclear concussion that would most assuredly have rendered their entire school into incinerated dust particles within a few seconds.

The same newsreel went on to portray how family members could protect themselves in the event of a nuclear attack. Huddled together under the work bench in the cellar awaiting the imminent detonation, they clutched their store of Campbell's soup, a flashlight and batteries, a few jars of home canned vegetables, a portable radio, and a first aid kit. The message conveyed was clear: The communists wanted to kill Americans, but with a little Yankee ingenuity a person could protect himself in the event of a nuclear war. Today it all sounds rather silly, yet in the prevailing atmosphere of the times, these newsreels were credible. They reinforced the message that the Soviets were enemies to be feared and they downplayed the real consequences of nuclear war. Now, however, much of our propaganda is more subtle.

One night in Leningrad, Molly, a friend of mine from our group, and I went out exploring by ourselves. We hopped on a bus in front of our hotel and set out for the center of town. We eventually ended up at the main bookstore where we decided to buy some posters. We happened to strike up a conversation with three young Soviet men and an American who had been living there for six months studying Russian. We began exploring the whole issue of how our governments use propaganda to serve their own needs. The Soviets had seen many recent American movies like *Rocky IV* and *Red Dawn*, courtesy of their ever-present black market. They considered these to be a powerful form of anti-Soviet propaganda. One young man made the point that American propaganda was in general more subtle than the blatantly obvious anti-American rhetoric he was used to in their own press.

I was a little surprised at their ensuing discussion that was openly critical of the Soviet system. I thought such discussions took place only in the most secret of places, certainly not in the middle of a public bookstore. It also astonished me that several of the men were openly trying to sell us black market lacquer boxes, a criminal offense. At the same time I was pleased that they so easily recognized each side's propaganda and knew that, in reality, we were probably little different from each other as people.

Sometimes when we hear something repeated often enough or long enough, we may no longer recognize it as inflammatory. In Kiev I asked the chairman of the local peace committee how propaganda by our respective governments contributed to the ongoing mistrust between our countries. He answered quite sincerely that he couldn't think of any instances of anti-American propaganda and would gladly give his salary to anyone who could point out any examples. I then told him about an English version of a Soviet newspaper I had just read at the airport. It used such inflammatory phrases as "Yankee imperialists," "capitalists," and "American aggressors" repeatedly, in writing about the United States. Blushing and obviously embarrassed, he conceded, apparently not realizing I was referring to something so basic. I didn't press him on the matter nor did I take him up on his salary offer.

In general, I found the Soviets quite well informed about American culture. They often knew more about American rock stars and politicians in Washington, for instance, than I did. At the same time, however, some Soviets I met had a lot of misconceptions about the workings of American society. A few wondered, for instance, if Americans all carried guns to protect themselves from rampant crime.

In the media of both countries there has been an underlying tone of prejudice that colors much of what is said. In the American media, for instance, subjects that reflect the negative side of the Soviet system have received a disproportionately large share of coverage. Stories about dissidents, refuseniks, injustices, the war in Afghanistan, and the relative lack of personal freedoms have predominated. There often has been an underlying tone of criticism, judgement, condemnation, and cynicism, as well as a general tendency to discount anything positive the Soviets have to offer. Proposals for peace coming from the Soviet government are apt to be spoken of skeptically or dismissed outright as mere "Soviet propaganda." Even *Time* and *Newsweek*, supposedly striving for objectivity, I have found to be biased at times, especially since I have returned from the Soviet Union.

During the last year, however, I have noticed a welcome softening of anti-Soviet rhetoric in American media, and a number of welldone documentaries on TV have shown Americans an unbiased view of Soviet society. This is a good example of a change brought about by the rising level of consciousness in the world—a trend that will continue and which will eventually make the widespread use of propaganda obsolete.

In contrast to my prior experience with American media, the Soviet literature I read seemed to have more coverage of positive events in the United States which helped to balance the negative. In the same paper condemning "imperialist war mongers" in the U.S. Government were articles about recent discoveries at the University of Kansas and news about the space shuttle flights. The Soviet press, on the other hand, is more apt to focus on American poverty, unemployment, inflation, crime, social inequality, racism, and militarism. Most Soviets I have talked to realize that what they read is often a distorted view of the U.S.

In both presses, the underlying techniques for discrediting the other side are similar. One journalist I saw recently on educational TV has studied propaganda from all over the world and concludes that all political cartoonists must have graduated from the same art school. They all portray the "enemy" as evil monsters with long sharp fangs, variations of the Devil, ruthless aggressors, or other stereotypical characters. The only difference is that the captions and labels are changed.

National prejudice is reinforced in two other major ways. The most visible and powerful of these is the political retoric spouted by our governments. Politicians, by their very nature, are masters at controlling public opinion. They also have instant access to national media coverage that guarantees their statements will be widely heard. When our leaders speak, many people unfortunately accept what they say as being the unbiased truth.

In the U.S., fear and mistrust can very easily be generated by military and political leaders intent on securing generous appropriations at budget time, by disseminating exaggerated ideas of "the communist menace." Soviets, because of a State-controlled media and a one-party system, hear only what their government wants them to hear. However, since glasnost Soviet viewers have been able to see indepth investigative reporting covering opposing sides of such controvercial issues as corruption, bureaucratic inefficiencies, and drug abuse. We like to think of "our" leaders as honest and that deception and corruption can only take place in "their" system. Watergate and the Iran-Contra scandal have made many Americans believe otherwise and some think that this is only the tip of the iceberg. In a like manner, Soviet citizens are beginning to demand more honesty from their leaders.

The other major way we generate the stereotypical characterization of each other as enemies is through our movies. Despite the obviously fictional plots in films such as *Red Dawn*, and *White Nights* they have the abiltiy to implant in Americans a subconscious apprehension and dread of the Soviets. Many Soviets I met had seen parts of the TV miniseries *Amerika* and quite understandably did not like the image portrayed of their country as the evil conquerors of the U.S. While Soviet audiences are also exposed to films with anti-American themes, they have a chance to view many more American-made films than their U.S. counterparts. Films such as *Kramer vs. Kramer* and *Convoy* with Kris Kristofferson were mentioned repeatedly by my Soviet friends.

A good illustration of how our preconceptions interfere with understanding one another could be seen in the series of space bridges in 1986-1987 hosted by Phil Donahue in the U.S. and Vladimir Pozner, a leading media analyst, in the Soviet Union. Via satellite Soviet and American audiences could see and question each other about day-to-day living in the other's country. Pozner is an exceptionally intelligent and articulate spokesman for the Soviets and spent much of his childhood growing up in New York before moving back to his homeland.

In the first show many in the American audience (including Donahue) were often critical, judgmental, and sometimes frankly rude when speaking to the Soviets. The speakers often conveyed a strong underlying prejudice against internal Soviet affairs of which they had no firsthand knowledge. The Soviets, I thought, showed exceptional patience towards the barrage of accusations. Donahue and Pozner finally were able to divert a potentially fruitless shouting match into a very productive exchange between the two groups.

As the space bridges progressed, however, there was a noticeable toning down of the recriminations and an increase in meaningful dialog between the two audiences. Both hosts are to be commended for their efforts in facilitating person-to-person contact. By the last show, with Donahue hosting from Moscow and Posner from New York, there was a definite sense of goodwill and friendship between the participants. Face-to-face communication like this is absolutely necessary to break down the walls of misconceptions we have of each other and establish the understanding that will bring us closer to peace between our peoples.

So how—and especially why—has this unhealthy rivalry come about? Are Americans and Soviets forever condemned to this dangerous and costly war of words because of vastly different political doctrines which can never be reconciled? George F. Kennan, a former ambassador to the U.S.S.R., traces the historical and underlying psychological reasons for our present disturbed relationship in an excellent book, *The Nuclear Delusion.* He concludes that humans have a subconscious need for an external enemy against whom their frustrations can be vented. We need to externalize our evil and create an enemy in whose alleged inhuman wickedness we can see the reflection of our own exceptional virtue.

Psychologists have labeled this "projection." Simply put, we despise in others the very qualities we can not accept in ourselves. This applies not only to societies and nations but to individual relationships as well. The next time you are critical of, or angry toward your spouse/friend/boss/etc., see if in all honesty the very thing you are intolerant of also exists in yourself, perhaps in a latent form. A good example is the person who campaigns vehemently against pornography because he or she is incapable of acknowledging his or her own subconscious erotic fantasies.

Indeed, if one traces the historical antecedents of the current U.S. - Soviet propaganda war, there was a time in our past when persecution of each other had all the qualities of outright racism. During the McCarthy era, for instance, to be an American meant having to be anti-communist as well. Any remote association with communism could have meant loss of your job without due process and other flagrant violations of personal freedoms guaranteed by the U.S. Constitution. If I had written this book in the Fifties, for instance, I could have been labeled a communist sympathizer and deprived of my right to practice medicine.

As an American, I uphold the principles upon which my country was founded. They are admirable ideals for all mankind, not just for people residing in the United States. One should be judged by his or her intrinsic worth as a person, not by race, color, creed, sex, national origin, religious belief, or political convictions. To condemn a whole nation because it supports communism or socialism or democracy or capitalism is unjust. Even those Americans who would condemn Communists have no reason to condemn all of the Soviet people when 90 percent of the population does not even belong to the communist party. And why condemn all Americans because of a few "imperialists" or "capitalists" who have exploited others for their own gain or because of a few "warmongers" who promote war for their own profit?

I highly doubt that the average American or Soviet individual has received an unbiased education about the other's political philosophy. I would wager that, if we put Soviets and Americans in the same room to talk about the principles they consider important in life there would be far more agreement than expected. We need to reexamine our ideologies and judge them not by blind adherence to a name, but by the intrinsic value they have when applied. As with anything else, the comparison between democracy and communism produces strong and weak points for each system—never a black and white answer.

What are the consequences of this propaganda war? By playing on our subconscious anxieties, the underlying mistrust, fear, suspicion—and sometimes outright hatred—fuels and perpetuates an ever-escalating and obscenely costly arms race. I was surprised at how real the fear of the United States was in some of the Soviet people I met. I was even more shocked when I learned of a recent American opinion poll that revealed that 54% of Americans "would rather be dead than Red." Both countries waste vast amounts of limited and vital resources that could otherwise be used for the betterment of everyone. Future historians will undoubtedly look back on this period as one of humanity's biggest examples of mass insanity and paranoia, a disease which so far has spanned nearly half a century.

To support this disease we have created a self-propagating international military-industrial complex and deeply entrenched government beaurocracies to foster its continued existence. All of us pay the price of keeping this monster alive. The few who do profit from it have, unfortunately, grown enormously powerful and undoubtably resist dismantling it. The world now spends close to a trillion dollars annually on the military in a time when glaring social needs go unfunded. Despite massive stockpiles of megatons, generals on both sides demand more or better bombs—even though we presently have enough to blow up every person on the planet with the equivalent of about four tons of TNT apiece. It is doubtful that any of this could have continued to exist without the underlying fears and mistrust nourished by decades of disseminated propaganda.

The roughly 10,000 nuclear missiles we have aimed at each other is much less a measure of our strength than it is of our insecurity. The $500 billion or so spent annually by the U.S. and U.S.S.R. for "defense" and "national security" buys us a world that is today much less secure than it was prior to the arms race. The defense establishments of both countries are powerless and impotent in protecting their citizens from a nuclear attack launched by the other side if one country choses to disregard the principle of Mutual Assured Destruction (M.A.D.). Today, rather than drawing out for years as past conventional wars have done, a full-scale conflict between the U.S. and Soviet Union would likely be over in a few minutes.

Somehow, there is a contradiction in building weapons of war to establish peace. As we have come to rely more and more on nuclear missiles, technology, and governments to preserve peace, we have given away our power to create peace and goodwill from within. Americans and Soviets both need to learn that it is not the fear of war that will maintain peace but a desire and commitment within each of us to create a climate of respect, cooperation, trust, and goodwill between our peoples. The more unconditional selflove and acceptance each of us develops, the closer we will be to global peace.

Although the effects of decades of brainwashing are still deeply rooted in our cultures, there has been a welcome reduction of verbal hostilities in the past few years. True, we still see examples of the old mentality, as in the recent series of expulsions from each other's embassies in 1986, but there has been a large number of recent TV programs that have given Americans the opportunity of seeing the Soviets from a much less biased viewpoint. Besides the Phil Donahue space bridges already mentioned, the ten-part PBS series *Comrades*, the ABC production *Seven Days in May,* as well as broadcasts of Soviet programs on American television have done a lot to dispell negative myths about the Soviets. As U.S. - Soviet relations improve further, there will be more widespread cultural exchanges and person to person contacts.

The myth that some people are inherently good and some are bad seems to be part of every culture's psychological profile. To be judgmental and racist is a potential human trait we all share. On the other hand, we all can be accepting, loving, and tolerant if we so choose. Jesus once told his followers to remove the log from their own eyes before attempting to remove the speck from their brother's. The problems we face in our own back yards are more than enough to consume all our efforts, without our trying to solve our neighbor's as well. If each of us begins by making peace within ourselves—the only part of the planet we really have control over—then the solutions to external problems will come easily.

To think that national boundaries can keep out human shortcomings, faults, and vices is a fallacy arising from self-ignorance. Examples of racism, persecution, brutality, drug abuse, corruption, the misuse of power, greed, and any other crime can be found on both sides of the Iron Curtain. None of us are immune because we all contain the seeds of all possible human behaviors. Given the right conditions and circumstances we, too, could become the very thing we despise in others. Only by acknowledging and accepting our frailties and forgiving ourselves can we move on to unconditional love of self and consequently of all humanity.

CHAPTER THREE

Face to face

Crossing the border into Russia from Finland was an emotional experience for me. Finally, after years of hearing about an imprisoned, mistreated, and persecuted people, I was actually going to see for myself who the Soviets really were. Standing outside the isolated border outpost in silence, I can remember looking for the first time through rising tears at Soviet soil and birch trees and swooping gulls from the nearby gulf of Finland. The brisk salty air and deep golden glow of the distant horizon looked the same as the land I once lived in as a boy. How could this be the same place I had been taught to fear?

Because the Finnish trains were on strike, we were transported by Finnish motorcoach from Helsinki to Vyborg, the first Russian city inside the border. There we disembarked in the early evening to catch an overnight train to Moscow. Since we had several hours to kill, some of the more adventuresome group members and myself decided to go exploring. Distant rock music and our bursting curiosity soon led us to a small bar and disco crowded with local Soviet young people out partying on a Saturday night. Timidly, at first, and then more boldly, as no one seemed to mind, we joined the tightly packed throng of oscillating bodies and danced to a universal language we all understood. To my surprise, some of the music was American. Before long, despite the language barrier, we began to make friends. This certainly was not the way I had envisioned our first Soviet contact, but it was typical of the rest of the trip. At that point, I knew we were going to have fun there.

As we left the disco to board our train, one of the young men who was either an off-duty soldier or draftee came out to see us off. In broken English (since we knew little Russian) he told us that he and his people wanted very much to be our friends and make peace between our countries. As we stood in a small circle communicating largely with gestures and eye contact broken only by tears, the bond of goodwill for each other generated in this spontaneous encounter was unmistakable. In that instant I saw how easily peace could occur if given a chance.

The next morning as I watched the speeding countryside and its small villages whisk by on our way to Moscow, I knew that I had a lot of learning to do about the Soviets and their land. Even its immense size is difficult to imagine. Wrapping nearly halfway around the world, the Soviet Union spans 11 time zones. As the sun rises in Uelen, the easternmost settlement on the Bering Strait, it is just setting in Kaliningrad, the westernmost city, on the Baltic sea. Within that vast expanse live more than 100 different ethnic groups speaking no less than 80 separate languages and writing in five different alphabets. President Reagan noted in his first summit meeting with Secretary Gorbachev that they both come from nations of immigrants, a fitting comparison as both countries represent a blend of greatly diverse cultures united under one flag. In that lies a common strength.

Despite an official policy of atheism, the Soviet Union's inhabitants continue to practice Islam, Judaism, Buddhism, and 152 sectarian versions of Christianity. No matter where we traveled, I was struck by the number of churches everywhere, a testimony of the deeply rooted spiritual heritage of these people. And wherever there wasn't a church, there seemed to be a war monument or memorial commemorating one of the countless invasions that have swept over Soviet soil for over 1000 years. Now that I know something of Soviet history, I can better understand why the peoples' desire for peace is genuine and deeply rooted. I don't think Americans can fathom the profound effect it must have to live in a land where every crossroad may have been a bloody battlefield in the past. Even though it ended over 40 years ago, World War II still affects the Soviet people today. It claimed the lives of 20,000,000 of their countrymen and left countless millions homeless and refugees in their own land. Many older people are still grieving their losses. Never again do they want war on their soil. Americans often mistake the Soviets' massive military strength as a sign of inherent aggressiveness and expansionist tendencies. In actuality, I think it reflects a peoples' will never to have their land overrun and destroyed by the ravages of war again.

I also often saw banners, signs, and murals proclaiming peace. Too often dismissed as propaganda by American officials, this visible appeal for global cooperation springs from the genuine longings of a people still paying the price for past devastation. I made instant friends and evoked heartfelt tears and hugs whenever I walked up to the rosy-cheeked Babuskas and spoke the words in Russian for "Peace and Friendship." (Interestingly the word "mir" in Russian means both "peace" and "world.")

Considering the vastly different historical backgrounds between Soviet and American cultures, it is not surprising that psychological differences have evolved into divergent ideologies. Springing from a largely feudal and aristocratic society only 70 years ago, the Soviets were much more likely to develop the collective viewpoint of communism than the American system of democracy. Americans value the freedom of the individual to create and live his or her own truth however different that may be from the view of the majority. The nonconformist and rebel have always found a niche in the American system. For the Soviets, however, security and conformity are perhaps valued more than individual freedoms.

At times I detected a definite sense of group consciousness among the Soviets that seemed absent from my own society by comparison. On the streetcars and busses there is an honor system for collecting fares. Coinboxes were out of view of the driver, yet everybody paid. When it was crowded, people passed the money hand to hand and the one closest deposited the coin and tore off the ticket. Many in our group felt safer walking on the streets of Moscow at night than in American cities, where even daylight may not discourage muggers.

One day several people in our group were walking past GUMS, the largest department store in Moscow. There, parked out front with no one in attendance, were several baby carriages with infants inside. Rather shocked, one American asked a passing pedestrian what would happen if something terrible occurred and the mother wasn't there. The woman answered simply that if the baby began crying then someone walking by would stop and rock the carriage until the mother returned. If the collective viewpoint has fostered this degree of group trust, then there is certainly something to be learned from it.

The Soviet state has been in existence for only 70 years, but it's built within a culture dating back over a millennium. Despite the state's desire to substitute its own ideology for certain traditional values, the influence of the past is unmistakable. For instance, having heard the word "Kremlin" often used in a negative context, I was surprised when I saw the place in person. Within the 70 acres encircled by its fortress walls are wellkept gardens, a marble theater where the Bolshoi ballet performs, and ancient gold-domed churches adjacent to the government buildings. Officially separate, their close proximity to each other bears testimony to a time when the church was a central feature of state authority. Perhaps in reaction to the church-dominated reign of Tsarist Russia, the present State in discrediting past religions has perhaps unconsciously substituted its own brand of secular religion, with Lenin as its patron saint.

Despite the official policy of atheism, the spirit of religion runs deep. Wherever I went churches abounded. Often in disuse, many have been magnificently restored at state expense. Regardless of your political convictions or religious beliefs, standing silently inside one of these ancient temples and looking at their frescoed domes a hundred feet over your head inspires a primal sense of awe. Some may call this feeling the presence of God, while others may explain it in more scientific terms.

Religious and other individual freedoms were much more in evidence there than I expected. Since religious and

St. Basil's Cathedral

Despite an official policy of atheism, architectual masterpieces like this symbolize a deeply ingrained tradition of spirituality in Soviet culture.

Kremlin Churches

To many westerners, the word “Kremlin” is the symbol of Soviet authority, often having ominous connotations. Here, church and state stand side by side, both making up the whole.

racial persecution are part of our human heritage, it is not surprising that examples of these can be found in the Soviet Union. I don't believe repression there today is as harsh as past well-publicized reports from dissidents have implied. Without question, rampant human injustices occured in the Stalin era and thereafter, and these undoubtedly have fueled Western anti-Soviet feelings since. The present Soviet leadership has shown a healthy attitude toward discussing past injustices and a willingness to allow open dissent, as evidenced by the recent release of Andrei Sakarov.

Constructive criticism of each other's systems is healthy, since we all at times are blind to our own shortcomings. Unfortunately, most of our past interchanges have consisted largely of blaming, condemning, accusing, and denouncing each other from an unrealistic platform of self-rightousness. It is important to avoid judging each other by our own set of ideals—which may be different from the other's. I found the Soviet constitution to aspire to many of the same human rights as the American Constitution. Article 52, for instance, guarantees each Soviet citizen the "freedom of conscience—the right to profess or not to profess any religion and to conduct religious worship." Thus, engaging the Soviets in a two-way dialog to facilitate the attainment of mutual ideals would be far more productive than hurling one-way verbal projectiles back and forth over the Iron Curtain.

While in Novgorod some of our group attended a Russian Orthodox Easter-eve service. Because there are no pews, the worshipers stand for the duration of the service which can go on all night. Despite conditions like these that would motivate many Americans to watch the ceremony on TV, churches are usually packed. The only form of religious

Babushka

The desire for peace is deeply rooted in women such as this one. They watched the majority of men in their generation perish in the nightmare of WWII that left 20 million Soviets dead. I would walk up to them and say the words in Russian for peace and friendship. They would take my hand and as tears welled up they would look me in the eye and say, "Da, da. . .mir (peace) mir" in return.

"I am but a poor man, but I am happy because God is in my heart." We often forget that freedom and joy can exist on the inside despite outward appearances or political climate.

Anatoli

repression I encountered was when the state-controlled TV network aired special popular programs the night before Easter to attract the attention of Soviet youth who might otherwise have been lured into church.

Because of Patricia's previous connections in Moscow, we were able to attend a Palm Sunday service at a Moscow Baptist church. The service had already begun when we arrived. People standing in the hallways and on the front steps, because there was no more room inside, graciously stood aside and beckoned us past them to the balcony where two rows of chairs had been reserved for us. Making our way as unobtrusively as possible to our seats, I noticed rows of people pressed against the sides and back of the church, standing, so there would be room for us to sit. The welcoming expressions on peoples' faces surrounding us put me at ease that we were indeed guests and not intruders.

As the service progressed, I could tell by peoples' expressions that they were deeply involved in their belief. Part way through the service—which until then had been conducted entirely in Russian—the minister turned towards our group in the balcony and in excellent English gave us a heartfelt welcome. From then on in an impromptu fashion, first one minister spoke to us in English, then another translated into Russian for the rest of the congregation. They had nothing but warmth and friendship to express, and said they were honored to share Palm Sunday with Americans. When the service was over and we got up to leave, there was warmth and love in the faces that looked at us and held our hands as we walked past. In that spirit of spontaneous brotherhood, I knew that these people and their countrymen deserved only my goodwill and friendship, not my suspicions and mistrust.

Dennis L. Haughton, M.D.

As our group gathered in front of the church to set off for the subway and our hotel, a man from the congregation in his late forties approached me. He introduced himself in halting, but very good English, saying his name was Anatoli. He had learned English from a book, only hearing it spoken a few times before. Since he was traveling in our general direction, he accompanied us on the subway back to our hotel. He was a simple man, radiated pride about his family, and was delighted to be able to practice his English with me. At one point he turned to me and told me something I will never forget. Opening his coat, so I could see the tattered cloth inside, he said, "I am but a poor man, I have very little possessions, but inside I'm happy because God is in my heart."

Anatoli opened my eyes a lot that day. He reminded me that what a man believes inside doesn't need the blessings or permission of a government to exist. Happiness has nothing to do with a man's material wealth. Poverty of material possessions will always be with us. Poverty of spirit will remain only as long as our ignorance prevails.

Governments and political ideologies come and go, but one thing never changes. At some point in life a person looks into the heavens or deep within and ponders the eternal questions that make life meaningful: Who am I? Why am I here? Who created this universe? Is there a God? What becomes of me when I die? Each will find his own answers, his own truth. For some the answers are found in organized religion in its hundreds of forms. For some the inward journey may reveal a more personal experience of the infinite life force or God within. Some may find nothing except what they can see and touch in their external space/time reality.

For others, devotion to political principles may give their life meaning. No matter what truth one lives his life by, it is nonetheless his own truth, even if it is not the truth of any other person. Governments or other forces may try to influence what truth we live by, but in our soul, in our heart, we are free to believe in anything we desire.

In the final analysis, it really doesn't matter where we live on this planet, what convictions and beliefs we claim as ours, what flag we swear allegiance to, or how rich or powerful we have become. When you come right down to it, we are all just people. With a few minor variations, we all live in the same body, use the same brain, think, feel, breathe the same air, drink the same water, and live on the same world. The Soviets I met seek the same from life as people the world over: to be happy; to meet their basic need of food, shelter, and clothing; to grow in skill, wisdom and knowledge; to come home after a day's work and enjoy the company of family, friends, or self; to gain more from life; and to live out their days in a peaceful world.

The Soviets I met in no way conformed to the stereotypical repressed, unhappy, and mistreated people pictured in my country's past propaganda. Far from parroting the party line, I found them intelligent and very capable of creative thinking on their own. We talked openly about the war in Afghanistan, human rights, our perception of their lack of individual freedoms, the war in Nicaragua, and many other sensitive topics.

They confronted us frankly about their perceptions of U.S. misdeeds. They seemed open to discuss the abuses of their past and imperfections of their present Socialist system. They readily acknowledged that they had not yet achieved their goal of a classless communist society. Most I

All Weather Ski Jump

Sports are a big part of Soviet life, often on a grand scale. This ski jump covered by a grass matting is open to all who dare. I watched 8 and 10 year olds fly off it one Sunday. Note the size of the people at the bottom. This jump is just the practice one for the big one pictured on the next page.

<u>The Big Jump</u>

talked to had seen vast improvements in the quality of life and had faith that the future would bring even more progress.

Like most people, I found the Soviets more concerned with the joys and problems of day to day living: working, raising kids, putting money away for something special, and ongoing relationships. Less then ten percent of the Soviets are communist party members, shunning the tempting lure of better privileges because the added responsibilities and obligations of leading an exemplary life are too demanding for most.

Despite official disclaimers, I have learned that there is a very definite class hierarchy in the Soviet Union. It determines what jobs and housing are available to a person, whether or not one can easily buy cars and foreign goods not available to most, and how easily a person can travel abroad. In the U.S., money buys power, privileges, and material possessions. In the Soviet Union, money is often not enough. The highest rewards and privileges, as well as salaries, go to important government officials and party members, sports or Bolshoi stars, and others, who for reason of their talents, have achieved national or international fame. As do people within any social system, the Soviets learn to live within theirs, and they find whatever shortcuts they can to better their lives.

Despite the totalitarian image portrayed in the west, the policemen and militia I saw there reminded me more of English bobbys than the cops I see in Arizona. Of course, Arizona is one of the few states where you can still see bikers driving by with holstered firearms legally strapped on their belts. Apparently the tanks and ICBM's pictured parading through Red Square that I have seen repeatedly on

American TV happened in the early 70's and have not been back since.

Once when I asked a group of university students in Kiev what they thought of the KGB, they replied that you just get to know how to spot them and stay out of their way. They also said that you soon learn how to get around laws that you don't like without getting caught. One night in Leningrad several of us went out dancing to a local disco that played 75% American rock music courtesy of MTV. By coincidence, we ran into two young men who were looking for a woman in our group who had ordered some black market caviar from them earlier. Offering to take us back to our hotel, one of them stood out in the busy street and flagged down a passing motorist who became an instant taxicab. Five Rubles quickly changed hands and soon we were standing in front of our hotel.

Now came the hard part. Since Soviet citizens were not allowed in our hotel, one of them borrowed a spare hotel card from us and the other decided to use his ingenuity and pose as a tourist who had misplaced his card. The scheme worked OK at the hotel entrance, but as we were attempting to slip unobtrusively into the bar on the top floor, the young man without ID was detained for further questioning. As his comrade joined us for a drink inside, he told us that the men who detained his friend were KGB. I asked how he could tell. He said simply that they stood out because of their clothes, their attitude, and their beady eyes. Sure enough, after a short time his friend was released and the agents came through the bar a little while later. Even I could have spotted them in the crowd. While everyone else was dressed casually, they were in suits and looked and behaved like plain-clothes cops do in the movies. I decided it was the beady eyes that gave them away. Then without further hassles, our friends found the lady who ordered the caviar, haggled over the price for at least an hour, and finally left the hotel with a bag of tapes, calculators, pens, and several other American items that got thrown in with the deal.

A basic trait we all share is the desire to be rewarded for our individual efforts. The Soviet system with its emphasis on central planning and collective equality has not yet fully utilized this attribute to its best advantage. It is not surprising that an outlet for this drive has been found in a widespread flourishing underground economy. In a society in which selling something for a personal profit is illegal, the existence of this vast black market is an embarrassment to the Soviet government. Some talented entrepreneurs have gone so far as building nationwide multimillion ruble empires that thrive on the sale of goods made from diverted state raw materials in state-run factories during off hours.

Of course, I never ran into these massive operations. Several times a day, however, I was approached by young men on the street, who either wanted to buy everything I was wearing, exchange money, or sell me caviar or other popular tourist items—at 30-50% of the price charged by the state-run beriozka stores catering only to foreign tourists. As one young man in Leningrad put it: The black market flourishes because of simple economics. He could earn far more in the underground economy than in a state job, so it was worth the legal risks. Doctors, for instance, earn an average salary of 145 rubles ($210) a month compared to the average per capita wage of 190 rubles ($275) per month. If a car costs 10,000 rubles ($14,500) and Japanese VCR's, always in short supply, cost between 3,000 to 5,000 rubles

($4300-7250), it is easy to see how tempting it would be to make a little money on the side.

The demand for foreign or American made clothing and other consumer goods is enormous. Most of the long lines I saw were not for basic necessities but for imported goods in short supply. I could have sold my Jordache jeans for 100 rubles ($145). One person offered me 250 rubles ($362) for my Casio watch that cost me $39 in the States. One young man I met in Kiev, Mikail, took me back to his flat (we had to elude a KGB agent by cutting through another apartment building) and was so in love with my new Nike running shoes that he offered me two bottles of Russian vodka, his own pair of leather shoes, a handful of peace pins, and a full sized Soviet flag in exchange. I felt guilty refusing his generosity. It was obvious to me he had his heart set on owning shoes like this someday, but somehow I thought it was best for me to abide by Soviet laws.

At present, certain limited forms of private enterprise are legal in the Soviet Union. Crops grown on a family's own small plot of land can be sold in farmer's markets in the cities. Not surprisingly, the average yield per acre on these private plots consistently surpasses the yield on adjacent, state-run communal farms. I predict that despite their fears of capitalism, the Soviet government will eventually see the potential benefits of free enterprise. Most likely, they will make it legal, call it something more acceptable to them, and as governments the world over are fond of doing, tax it heavily. That way everybody gains. (Indeed, since the preceeding was written, the Soviets, following the 27th Party Congress in March, 1987, have adopted some sweeping changes in an effort to revitalize their sluggish economy, declaring that if certain capitalistic ideas are good, they have

Wendy

As a professional video artist, Wendy recorded all our experiences on camera. Like children everywhere, Soviet kids were delighted to see themselves on instant replay.

Young Pioneers

We ran into these youngsters in Red Square and instantly fell in love. Look at the wonderment radiating from the girl in red. Their attention has been captivated by a professional clown in our group that delighted kids everywhere.

Romni and Friends

Despite the language barrier, we communicated well with each other nonverbally. Above, Romni makes a finger puppet mouse come alive—seen with the boy at right—for a group of spellbound Pioneers.

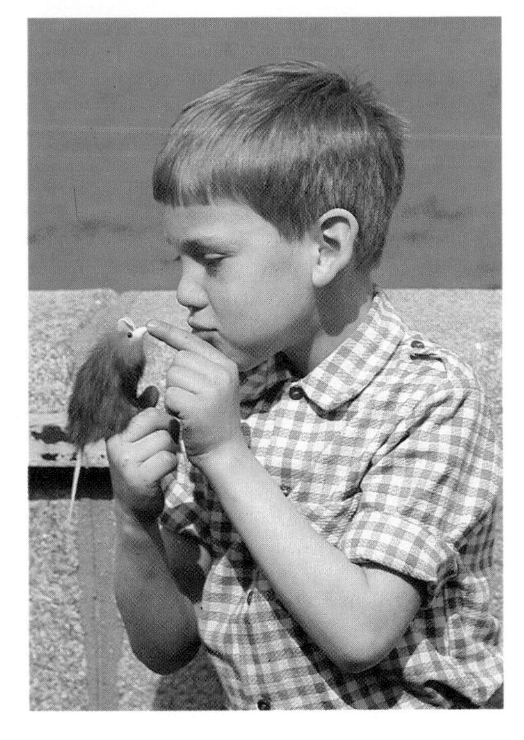

no ideological hesitancy in using them. Now, many forms of private enterprise from running a car repair business to operating a private restaurant are becoming legal.)

I never met anyone in my travels who was hostile, rude, or unfriendly because I was an American. On the contrary, most of the people I met were fascinated with the U.S. Often they had more knowledge of American geography, history, political policy, and rock stars, than I did. Most hoped to visit the United States someday. Only one I met, a student in Leningrad, wanted to leave the Soviet Union permanently.

I had heard from others who had toured the U.S.S.R. that they didn't see many smiles but I saw plenty. One night returning to my hotel in Moscow after dark, I was surprised to find singing and dancing in front of the Metro entrance. A Gypsy-looking couple danced on the sidewalk accompanied by a fiddler and accordion player from the sidelines while a clapping crowd gathered in a circle around them. I found myself thinking: this sort of thing isn't supposed to happen here. Before long, some people in my group started dancing in the circle also and soon a crowd of more than a hundred were caught up in the celebration. I finally decided I wanted to eat more than I wanted to clap and sing so I crossed the street to the hotel before the party ended.

In general, the standard of living in the Soviet Union is much lower than in the U.S., where we have been spoiled by abundant material wealth. Some Americans, used to color TV's in every room, a swimming pool in the back yard, and a Mercedes in the driveway would have trouble adjusting. Cars in the Soviet Union may easily cost three to four years' wages and years on a waiting list. Japanese VCR's are often not available at any cost. Some people may have to wait months or years to get a new apartment. A Soviet shopper visiting an American supermarket for the first time might very well pass out with disbelief. Shopping Soviet style I found meant standing in one line to see what is available, moving to another to pay for it, and finally waiting in a third to pick it up with your receipt. Americans used to choosing between several dozen brands of toilet paper in different hues and scents to offer their bottoms might easily develop blisters from the Soviet version, when it's available. What American soda drinker, accustomed to sipping on perhaps 20 different varieties of cola alone, would be satisfied with the only brand I ever encountered—regular Pepsi in reusable bottles?

Does infinite selection of consumer products make one society happier than the other? Freer? Superior? Probably not. On the other side of the coin, you can ride anywhere in any Soviet city comfortably, quickly, and safely for five kopecs (about seven cents). Monthly rent, utilities, and phone bills may only cost 18 rubles ($26)—total. The Soviet Constitution guarantees every citizen a 41 hour maximum work week, 100% healthcare, a home to live in, free education including college, paid vacations, and complete disability coverage and maintenance in old age.

There are pros and cons to both the American and Soviet systems. The simplistic condemnation of one system by the other is becoming less acceptable as people on both sides become more enlightened. Whereas a comprehensive comparison of the two societies is beyond the scope of this book, several points are worth thinking about. When Americans criticize Soviets for their relative lack of individual freedoms, Soviets answer that often they perceive

Americans abusing their freedoms. They cite the widespread drug problem in America as an example. In addition they note that the rampant epidemic of AIDS in the U.S. is a clear consequence of sexual promiscuity. Indeed, some Soviets are skeptical that Americans can even exist in a society without the familiar controls and restrictions accepted by them as necessary parts of life. To Americans, Soviet-style police tactics seem offensive. To the Soviets, the American legal system is criticized for letting criminals and killers escape punishment and roam free. Americans point out the often crowded living conditions in the U.S.S.R., with apartment-sharing seen as a disadvantage. The Soviets point out the millions of homeless Americans living in the streets and the often impoverished and abandoned older generation. The comparisons and contrasts are endless, the conclusions rarely black and white.

Without doubt, our preconceptions of each other are an unjust distortion of reality that has perpetuated the Iron Curtain of misunderstanding between our peoples for the last 40 years. What I have seen and experienced first hand in my travels has given me a refreshing vision of our future that was impossible in the light of old dogmas and limited thinking. The U.S. and the Soviet Union have much to learn from each other's experience of life and much more to gain as friends than as antagonists. As people of this world we have far more in common than the differences which seem to separate us. Once we are willing to see each other face to face instead of through our distorted misconceptions, the walls that keep us apart will quickly fall.

Moscow from Hotel Cosmos - 28 April, 1986

Unknown to us and the rest of the world, reactor #4 at Chernobyl to our west was bruning and we were headed for Kiev—only sixty miles away from it.

CHAPTER FOUR

Chernobyl

Midway through my trip in the Soviet Union, an event of global significance occurred that shocked the world into recognizing the deadly consequences of nuclear science turned against us, whether we wanted to see it or not. The following is a day by day account of that disaster as it unfolded.

Friday April 25, 1986

Chernobyl:

It was the beginning of what looked like a beautiful spring weekend at the Chernobyl power complex. Situated a little over 60 miles north of Kiev on the Pripyat river, the four reactors in current operation plus two more scheduled for completion in 1988 would make Chernobyl the biggest power station in the world. Its output of 6000 megawatts would be enough to light up all the houses in England. Nearby were the small towns of Lelev, Pripyat, and Chernobyl where many of the power plants workers lived and raised their families.

In all, about 100,000 people lived within 30 kilometers of the plant. Many would take the advantage of the warm spring weather to enjoy outings or fishing trips in the surrounding countryside. The May 1st holiday was rapidly approaching and some would use this weekend to prepare for the upcoming celebrations. No one in their worst nightmares imagined the hell that was to break loose in a few hours and change their lives forever.

Inside unit number four, technicians were preparing to shut down the reactor for its annual maintenance. Plans had been made to use this opportunity to run an experiment. The operators wanted to see how long the station's steam turbines could generate power for supplying emergency systems after the steam supply was shut off and before the emergency diesel generators had time to start. Although the plans had been submitted to the designers of the station, they never were approved and authorization for the experiment had not been given.

1 A.M. The process of shutting down the reactor began as control rods were lowered into the core. Over the next few hours the technicians planned to slowly reduce the reactor's thermal energy level from the normal operating level of 3200 MW, to between 700-1000 MW set for the experiment.

2 P.M. Because the emergency cooling system would draw power and affect the test it was shut off at this time. This was the first of six fateful violations of safety procedures that were made that night. Had only one of them

been omitted, the accident wouldn't have happened. Reduction of power was halted for about nine hours when an urgent call from the local grid controller in Kiev indicated that unit four's power output was needed to meet demands for several more hours. The emergency cooling system, however, remained disconnected.

* * * * *

Saturday April 26, 1986

Chernobyl:

12:28 A.M. Now the operators committed their second critical error. A regulator was set incorrectly and power output plummeted to a dangerously low 30 MW, far too low to continue the experiment. At this point, rather than abandoning the test which couldn't be repeated for another year when the reactor was again shut down for maintenance, the operators pressed forward. To increase the reactor's power output they began frantically withdrawing control rods until only 6-8 were left in the core. This was their third critical violation of safety rules. Standard operating procedure specified that no less than 30 control rods (occasionally 15) were to remain in the core at any one time.

1:00 A.M. Power levels rose and stabilized at 200 MW briefly but were still too low for the experiment.

1:03 A.M. In order to help circulate cooling water through the reactor's core, operators turned on two extra pumps. This created a highly unstable balance between water and steam levels in the system that demanded frequent second-by-second manual valve adjustments. This was their fourth mistake since it caused the system to behave unpredictably.

1:20 A.M. Here, for reasons which are unclear, they committed their fifth violation of operating rules and switched off an automatic shutdown system that would have stopped the runaway reactor had it been in operation.

1:23 A.M. Finally, they began the experiment and made the last of six fateful errors by turning off the only remaining safety system that could have prevented the disaster from occurring less than a minute later. At this point they had, in effect, turned the reactor loose to do as it pleased, with most of the control rods out and all safety systems disconnected.

1:23:30 A.M. By now, the control room supervisor knew something was seriously wrong and ordered all control rods to be driven back into the pile. Unfortunately, by this time it was too late. Four seconds later the reactor's power surged to 100 times its normal capacity, part of the core went critical and an explosion equal to 1/2 ton of TNT blew apart the reactor. Contact of the zirconium cladding with the cooling water and the subsequent production of hydrogen caused a second massive explosion several seconds later. This tore the reactor's 1000 ton lid off and brought the 200 ton refueling crane down on top of the reactor, damaging more cooling circuits. As fragments of fuel and burning graphite were ejected through the breached containment structure, 30 different fires were started in the surrounding complex. Only the suicidal bravery of the firefighters that night prevented the fires from engulfing the adjacent reactor number three. With nothing to stop it now, a plume of deadly-poisonous fission products rose high into the atmosphere on flames estimated to be 500 meters high and thus began its fateful journey around the globe that would affect the lives of millions.

* * * * *

Sunday April 27, 1987

Sweden:

2:00 P.M. The cloud, which initially rose to an altitude of nearly six miles, now had traveled silently over a thousand miles northwest and crossed the border of Sweden. Although unmanned radiation monitors along the Swedish perimeter recorded a rise in radioactivity, it wasn't enough to set off any alarms. Except for the people intimately caught up in the containment and evacuation of the Chernobyl area, plus members of the Soviet government, the world was still ignorant of the poisonous vapors looming over its head.

* * * * *

Dennis L. Haughton, M.D.

Kiev - 28 April, 1986

By any standard, Kiev is indeed a beautiful city. Invisible in this picture and unknown to me at the time, a deadly plume of poisons is surging from reactor #4 and spreading its vile mist on an unsuspecting world.

Monday April 28, 1986

Sweden:

9:00 A.M. At the Swedish Forstmark nuclear power station an arriving worker set off an alarm when he stuck his feet into a radiation detector. Since he had not been in a radioactive area, the plant managers at first concluded that a leak had developed in their own reactor. Fearing the worst, the plant was evacuated, the roads barricaded, radio warnings were sent to the surrounding community, and a frantic search was made for the source of the leak. When none was found, the rest of the plant's workers were scanned with Geiger counters. Readings of 5-10 times normal background radiation on everyone's clothing meant that the source of the contamination was outside the plant.

Other areas in Sweden and Finland also began registering increased levels of radiation. An analysis of the weather patterns over the previous two days indicated that the source had to be within the Soviet Union. Since there had been no seismic activity suggesting escaping radioactivity from a nuclear weapon's test, there was only one logical conclusion: there must have been an accident at a Soviet nuclear power plant. For the remainder of that Monday, Swedish inquiries through diplomatic channels got no information from the Soviets.

Kiev:

Afternoon. In the early afternoon, suspecting nothing of the burning power station 60 miles to the north, my group landed in Kiev on a jet from Moscow. Having spent the last few days in a crowded city, the lush spring greenery and newly erupted bright flowers in the countryside was a welcome sight. Kiev is indeed a beautiful city. Sprawling green parks along the Dnieper river, stately black marble buildings, well-kept cobble-stone streets lined with chestnut trees, impressive monuments, refreshing fountains, and bountiful gardens of bright red tulips are among the images that drift into my memory as I recall Kiev.

We spent the afternoon leisurely exploring and relaxing. Many in the group were tired from travel. By evening a large number of us had the first symptoms of food poisoning. A cold salad with eggs and mayonnaise had sat on the tables of our Moscow hotel several hours too long awaiting our delayed return from the Baptist church service. I only ate a little of it myself and therefore wasn't very sick, but some others were not as lucky. Most people recovered in 24 hours, some stayed in bed the next day. However, Patricia's son, Eric, got so sick he needed the services of Soviet physicians plus intravenous fluids administered in his hotel room.

Evening. That night, much to our dismay, our generous Ukrainian hosts served us a hearty five course feast. By this time the mere sight of food caused some people to turn green and lurch toward the restrooms. As we politely picked at our food and ate what we could, our hosts must have thought we were just spoiled and fussy American tourists. They carried much of the food, uneaten, back to the kitchen. In our fragmentary Russian and sign language I think we finally were able to get the truth across.

Still there had been no official announcement of the explosion at Chernobyl. At supper however, Patricia drew me aside and told me quietly that there was a rumor going around regarding some kind of nuclear accident, but not to tell anyone yet as it might be unfounded. I remembered her telling me at LA International airport as we were departing the U.S. that she had a premonition of some calamity that would disrupt our tour, although none of us would be seriously hurt. That night I didn't give it much more thought before retiring early to bed.

<u>Moscow:</u>

<u>9:02 P.M.</u> Finally the official Soviet announcement about Chernobyl came. On the evening news program *Vremya* the following statement was read:

"An accident has taken place at the Chernobyl power station, and one of the reactors was damaged. Measures are being taken to eliminate the consequences of the accident. Those affected by it are being given assistance. A government commission has been set up."

That was it. None of us even heard of the announcement until the following day.

* * * * *

Tuesday April 29, 1986

Kiev:

The next day business in Kiev went on as usual. No panic. No anxious glances at the sky. No sirens or firetrucks. Public transportation seemed to be moving people to and fro in an orderly fashion, despite the recent unannounced dispatch of 1,100 busses sent quietly northward to help evacuate the towns and villages around the burning reactor. Our group spent the morning sightseeing. The magnificently frescoed ceilings of St. Sophie's Cathedral built in the 11th century rendered me speechless. Talk of Chernobyl was nonexistant since news hadn't filtered down to us yet.

Then abruptly, the shit hit the fan. I was relaxing in my hotel room after lunch when suddenly the phone rang. It was Marti, one of our tour leaders, telling me to come quickly as I had a call waiting from the State Department in Washington. I, of course, thought the worst. Nobody gets calls from Washington while overseas unless something terrible has happened. Bracing myself for the possible news that my mother or son had died in a freak accident, I leaped up the stairs and raced into Marti's room.

I was somewhat relieved when the woman on the phone reassured me that my mother was fine, but that she was frantically trying to find out if I was safe. I told the woman that as far as I knew I felt well, and asked if this had anything to do with the rumor I had heard the preceding evening about a nuclear accident. At that point the woman became very serious and related to me some of the news that had appeared in American media since the Soviet announcement Monday night: The reactor at Chernobyl had indeed blown up. Early reports indicated that 2000 people had been killed. The area around the reactor site had been devastated and Kiev was being evacuated due to intense radioactive fallout. I now knew why my mother had reason to believe I was dead. When I informed the woman that there were no signs of panic or evacuation efforts she sounded greatly relieved.

By early afternoon other members of my group had received panicked calls from home, relaying gory details of the accident that fed our rising anxieties. Patricia was able to get bits and pieces of information from Soviet and American contacts, so that by the end of the day we were able to piece together a bare outline of what was going on. We knew that 80 kilometers away a nuclear reactor had exploded and was still burning. We were told that prevailing winds were blowing north, away from Kiev, which meant that we probably hadn't received any fallout yet. We had listened to the stories relayed from the States. That is all we had to go on for the next few days.

Needless to say, many of us were more than a little worried. Lack of concrete information allowed our imaginations to supply the possible consequences. The fear that we could all die together brought us closer as a group and brought down walls that otherwise may have never been breached. To this day, I still feel a close bond with those people and with the people of Kiev. A few in the height of panic would have left immediately for home had it been possible. Some in our group were angry at the Soviet government for allowing us to fly into Kiev in the first place. As fate would have it, our scheduled departure from Kiev on Wednesday was changed to Thursday afternoon. This is not uncommon in the Soviet Union. Whether it had anything to do with the disaster, we will never know. It did cause further anxiety, however.

Most of us had little preparation for what to do in a nuclear disaster. This probably reflects the general level of ignorance about such matters in the United States. In an act of moral schizophrenia, perhaps we have to repress the details of nuclear catastrophe along with our anxieties about the bomb in order to continue expanding our nuclear arsenals. I was surprised that the Soviet school children had been taught in graphic detail the effect of a nuclear device detonating over a city. In all my medical education, I never had any training in dealing with widespread radioactive fallout. Fortunately Patricia had some knowledge in this field and advised us to stay indoors, limit our excursions in Kiev, keep windows closed, drink bottled water, and avoid eating leafy vegetables. Largely because of my own ignorance or good job of blocking out my fears, my anxiety level was fairly low and I continued to enjoy the sights and people of Kiev in the streets. I deluded myself into thinking that whatever was happening 80 kilometers away couldn't possibly affect us. We later would find out that Patricia's advice was wise. Those of us who spent the most time outside had the highest radiation readings.

During our entire ordeal in Kiev, Patricia's spiritual guidance kept our energy focused on healing rather than fear. Out of turmoil she created a growing experience. Where some can only see problems and obstacles, she sees stepping stones to growth. She was able to show how the wisdom gained by the world from this seeming cataclysm would hasten the healing of the nuclear madness that still had mankind in its grip. Without the gift of Chernobyl, mankind could continue to deny the consequences of a runaway nuclear arms race in a divided world where peace was based on the threat of mutual destruction. Chernobyl was a harsh reminder of the effects of even a small nuclear war. Apparently, we needed to see that.

* * * * *

Wednesday April 30, 1986

The United States:

The news vacuum produced by the Soviet's decision not to provide timely information about the disaster was quickly filled by the imaginations and speculations of the western media. Without access to the details for on-the-spot coverage, the media depended solely on information gathered by both military and civilian reconnaissance satellites. Since the images available of the Chernobyl complex were open to differing interpretations, the speculations generated by them vastly distorted the real nature of the accident. A UPI story Tuesday evening had reported that 2,000 had already died. A headline in *The New York Post* proclaimed 15,000 had been bulldozed into a mass grave. Two ham radio operators with a sick sense of humor added further momentum to rumors already out of control. They falsely reported interception of a transmission from the Kiev region saying that reactor three had also melted down. This distortion was fueled further by Pentagon sources claiming that the fire in unit four was getting worse and that fire had possibly spread to the adjacent reactor.

These rampant speculations were in direct contradiction to official Soviet releases claiming only two had died and another 197 injured. Western governments and media were quick to condemn the Soviets for being secretive and withholding details: a seeming violation of Gorbachev's new policy of glasnost, or openness. The media and many western governments seemed to be turning the Chernobyl misfortune into anti-Soviet political rhetoric. As it later turned out, the Soviet casualty figures were correct all along. Later in a speech on Soviet TV, Gorbachev quoted some of American media's most outrageous speculations then denounced them as "a real heap of lies." He then went on to a little bit of anti-American rhetoric of his own.

Kiev:

Despite our misfortune of being in a disaster area, we continued to go about our activities in Kiev without too much interruption. I was there to meet Soviet people and soon had several new Ukrainian friends. As long as we were told that the prevailing winds were in our favor, I decided that going out was worth the chance. The people of Kiev seemed particularly friendly.

On a more official note, we met with the local Peace Committee. Supported by private donations and staffed largely by volunteers, the committee's purpose was to advance the cause of world peace through education, public awareness campaigns, and peace rallies. The chairman was proud to show us pictures of a recent rally involving over 100,000 participants.

I found the desire for peace to be an integral part of Soviet social consciousness. Many of the banners and slogans I saw draped and painted on buildings depicted peace themes. Particularly interesting to me was an article in the

Hippocratic oath affirmed by new Soviet physicians in which they pledged to do everything in their power to prevent nuclear war. At no time did I get the impression that people were simply regurgitating the party line about peace. Their desires seemed genuine. In fact, spontaneous expressions of goodwill toward us occurred so often in Kiev that it became impossible to believe that these gentle people were considered enemies by some back home.

Everything seemed normal that night when I went to bed. The sky was still clear although a noticeable haze had developed. None of the city's residents seemed overly anxious. Certainly no one showed any signs of panic. I still didn't know the magnitude of the raging fire 60 miles away that continued to spew its deadly toxins into the night. Despite the warm night I reluctantly closed the window. Feeling comforted in the belief that the wind was still in our favor, I nodded off to sleep. Little did I know, however, that the wind had already shifted and Kiev was beginning to receive a silent rain of fallout.

* * * * *

Thursday May 1, 1986

Kiev:

The sun shone brightly that morning and created a perfect day for a holiday parade. Looking anxiously at the sky I saw no ominous dark smoke, no falling cinders. Falsely reassured, I prepared my camera gear for the festivities. Some had decided to heed Patricia's warnings and watch the parade from the safety of their hotel room but when I couldn't

May Day Parade - 1 May, 1986

This holiday had its beginnings in Chicago in the late 1800's. Winds had now shifted and were spreading a rain of deadly fallout from Chernobyl on everyone outside that day.

People of Kiev

Smiles and happy faces were everywhere. Like the flower children of the 60's, people walked up to us on the sidelines, smiled, and handed us flowers.

find the right camera angle, I decided to risk being on the street. Sounds of clashing symbols and beating drums echoed from the distance as bands warmed up. The militia were beginning to form an honor guard every 10-20 feet along the parade route. School kids in bright uniforms and mothers with children in their Sunday best scurried across the square in front of our hotel to reach their assigned places in time. In Kiev, it seemed, you don't watch a parade, you march in it.

May Day in the Soviet Union is similar in importance to July 4th in America. A four day holiday originating in the labor movement in Chicago in the late nineteenth century, it is celebrated widely throughout Europe. In retrospect, I am glad we were in Kiev and not Moscow for the celebration. Here we enjoyed an intimacy with the participants that we may not have found in the more formal observance at the capital. There for instance, only a select few are privileged enough to watch the parade from Red Square, certainly not ordinary citizens or low-ranking American visitors. Although we had permission to observe the procession near the central square, I chose the less crowded area in front of our hotel. Pedestrian movement along the sidelines was restricted by multiple check-points requiring proper ID to pass. Once the parade started, all were required to stay at their assigned locations.

Contrary to expectations, there were no tanks, rockets, guns, or squadrons of goose-stepping soldiers in this parade. Many of the floats promoted world peace. Banners of Lenin and other Soviet heros abounded, but mostly there was wave after wave of people, each group in their own distinctive costume: women in traditional brightly embroidered Ukrainian dresses; aging war veterans heavily

Romni and Admirers

By the end of the parade many American onlookers had an arm's load of flowers that marchers had given them.

Ukrainian Man

Somewhere in a silent silo in Kansas there is a nuclear missle aimed at this man and his friends. Everywhere I went, Soviet people expressed nothing but goodwill and peace for Americans. They, too, tire of the political struggle that has separated us and want Americans to know that they want to make peace.

laden with medals; young Pioneers in their red and white outfits; sports teams including one wheeling their soccer goal behind them; school bands each in distinctive uniforms; and of course, thousands of ordinary citizens marching in their best attire. It was a day of celebration, people were happy and smiles were everywhere.

In preparation for this event, the schoolchildren had made thousands of flowers for the marchers to carry. Brightly colored paper of many hues, sticks from trees, pieces of wire, and a little glue were transformed into blossoms that I thought were real until I looked more closely. All during the parade while we were standing along the curb, people would walk up to us, smile, hand us a flower and then continue on. I was reminded of the flower children in the U.S. from the 60's. The spirit of good will was so strong I found tears welling up inside as I watched. By the end of the parade, several women in our group had their arms full of gifts.

The exchange of gifts and friendship was a two-way street. Wendy, our video expert, and another woman had brought their Polaroids and what seemed like a suitcase of film. As marchers passed, they would snap a picture of someone and then run out and hand it to the person. From the looks of surprise I saw, I guessed that Polaroid was rare in the Soviet Union. Puzzled at first with the blank print, people later burst into smiles of astonishment further down the street as their images began to appear. Many came back after the parade and asked to have their picture taken with this wonder of American technology. Since it was impossible to buy good quality Western film there and I heard that Soviet film was unreliable, I panicked to find that my own supply was dwindling rapidly. Luckily, however, by begging and bribing, I was able to scrounge up a few extra rolls from others in the group to last through the following week.

After the marching had stopped, I decided to go out exploring on my own. Again, I had a false sense of security. I assumed that authorities would have issued warnings if the city were getting fallout. So, against Patricia's advice, I hiked towards the heart of the city. When I got to the central plaza I was glad that I had come. About six to eight large circular fountains and lots of bright flower gardens were spread out in an open mall built for pedestrians with plenty of benches for sitting and relaxing. Families in their holiday dress were parading about as freelance photographers were selling their services.

What caught my eye immediately were the kids playing in the fountains. As proud parents looked on with beaming faces from the sidelines, the children waded in the pools and threw their balloons into the spurting jets where they were carried aloft to the sounds of their squeals of glee. It was love at first sight and I spent the next several hours attempting to capture it all on film. Little did I know at the time, but all that day a silent, invisible, and toxic rain of fallout was descending on us from the burning reactor 60 miles away. Playing in the water on the dusty ground where radioactive fallout is concentrated, these kids in their innocent jubilation may someday pay the price for our nuclear mistakes.

Later that day, as I boarded our bus for the airport, I had mixed feelings. I was glad to be leaving the immediate vicinity of Chernobyl, to be sure. Still, we knew only the barest minimum of facts of what had happened. At this time, we didn't even know if Kiev had gotten any fallout, and if so,

Patricia

She was the spiritual force that guided us through the nightmare of Chernobyl. Despite everything, she radiated love thoughout.

how dangerous it was for us. Anxiety levels ranged from minimum to extreme. The slightest little symptom of headache, nausea, or pain evoked fears of impending radiation sickness and death. While the outside world was inundated with conflicting coverage of the reactor explosion, we, without access to any western media, knew only a few bits and pieces of news from our Soviet guides.

Halfway to the airport, a sudden explosion from a blowout rocked the bus to a halt. The location couldn't have been chosen better. We were stranded in the country by a roadside rest area and there was a pathway which led directly into the cool forest. Accepting the invitation, we welcomed this quiet interlude far away from the city bustle to go exploring and to contemplate our fate. Like all forests, the smells and sounds of nature all around brought renewed peace for the next hour as we awaited another bus.

Before long we were airborne and headed back to Moscow, putting welcome distance between us and the still burning reactor. In retrospect, it probably was good that we couldn't buy *Time* or *Newsweek* in Kiev. The frenzied speculations circulating in American and Western media would have just intensified our own fears. Anxiety and stress have a profound effect on our immune defenses. Through futile worry, we could have hindered the process of inward spiritual healing through which Patricia was leading us.

I will always feel a deep bond with the people of Kiev because of what we shared together. Whatever happens to the world affects us all, for we all ride the same planet through the heavens. There must have been a reason for us to have been there then as citizen diplomats from the United

States. The lessons of Chernobyl were not only meant for the Soviets. Its wisdom was given to us all. Someday, I plan to revisit Kiev when I can be sure that radiation levels are safe. But, for the moment I am content to travel there in my memory to visit the friends I made on those warm spring days last year.

CHAPTER FIVE

Out of the inferno and home again

I was glad to get back to Moscow and put the ordeal of Kiev behind me. As we walked through the city Thursday night watching the fireworks display that signaled the end of May Day, I felt good to breathe air that I was reasonably assured didn't contain bits and pieces of a burning nuclear power plant.

Just as Patricia had envisioned before we left the U.S., our itinerary was disrupted by the accident. Because of our delay in Kiev, we missed our scheduled meeting with the Moscow Peace Committee, the U.S. - Canadian Institute, and the editors of the Soviet Women's magazine. I was a little disappointed to learn Friday morning that the Soviets had requested that we be taken to a local Moscow hospital to be checked for radiation instead of going on our planned tour of St. Basil's Cathedral and the Kremlin.

Actually, at this point, we weren't sure if we had even gotten *any* radiation, let alone how much. The information available to us on Chernobyl was scant and we had no knowledge at all of radiation levels for Kiev and the surrounding countryside. Most of us still felt well, although we

Moscow Hospital

We were taken to this hospital on our return to Moscow to be checked for fallout contamination. Below, technicians scan our bus for signs of radiation but find none since it originated in Finland and not Kiev.

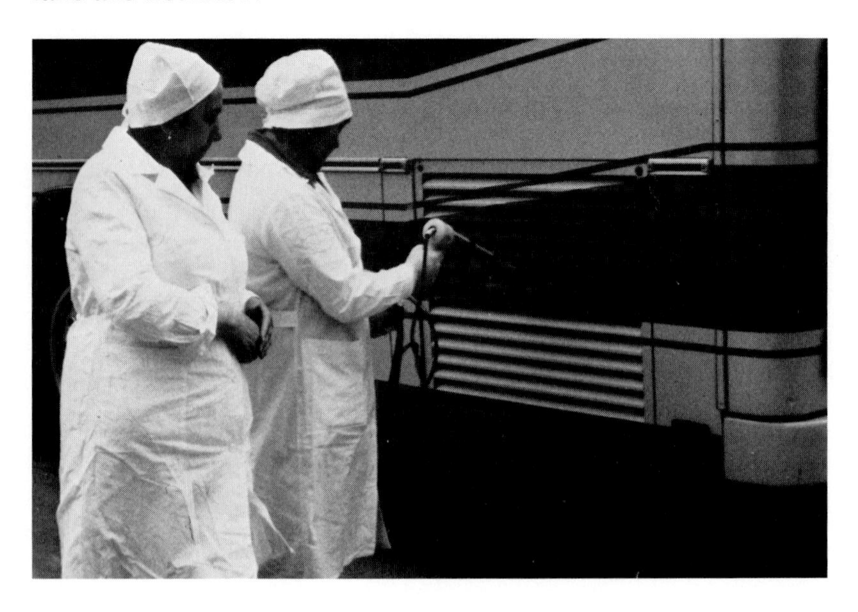

continued to joke nervously about the slightest symptoms as though they were the onset of radiation sickness. I preferred to remain in blissful ignorance. After all, I reasoned, even if we had been dosed with fallout, very little could be done after the fact.

We spent a good portion of Friday at a Moscow hospital and here we learned that we definitely were radioactive. We were led into the hospital in small groups. I was taken into an examination room and scanned very thoroughly with Geiger counters from head to toe. Although I later learned that their equipment was archaic by U.S. standards, the procedures they used were as systematic, methodical, and professional as any I saw in the States when I returned. As various parts of my body were scanned, I could very clearly see the needle on the instrument deflect out the normal range. I watched anxiously as some peace pins on my jacket caused the needle to fly off the scale completely. Only by switching to progressively less sensitive scales, could they obtain a reading.

By now I knew I was in trouble. I tried to persuade my racing heart to slow down as I contemplated an imminent death. As various hot spots on me were discovered, one technician typed the figures onto a medical report, while another recorded them in a book. Unfortunately, since the numbers and measuring system were in Russian, I couldn't interpret what they meant. Even had they been in English I wouldn't have understood their significance since basic nuclear medicine was not part of my medical training. It was bad enough to see the Geiger counter needle jump when pointed in my direction. One didn't need any education at all to know that human beings are not ordinarily radioactive.

I was next asked to remove my shirt and a doctor listened to my heart and lungs and performed a brief physical examination. Then a laboratory technician took a drop of blood from my finger for analysis and had me give a urine specimen. I was a little shocked to see the same lance used in drawing my blood wiped in alcohol and then used again on the next patient. (I was reminded of getting blood drawn for a marriage license in Massachusetts in 1966. After drawing blood from my arm, the 70 year old G.P. washed the needle and syringe under tap water and proceeded to take blood from my fiancee.)

Finally, at the conclusion of my exam, the doctor gave me a typewritten copy of my medical report in Russian. He then explained to me in English that the radiation levels detected on me were not dangerous. I was then discharged without any precautionary instructions and I raced back to the bus to compare notes with other group members. Some had little or no detectable readings. Some, especially those who had spent considerable time outside in Kiev, had much higher readings. To them, the Soviet doctors recommended washing or discarding the more highly contaminated items. While this advice seemed logical and was repeated to us many times by American experts upon returning home, I later found that it was impossible to decontaminate clothing by washing.

We now knew who was radioactive and who was not. We did not know what this meant. Despite Soviet reassurances that the levels were safe, many of us were still skeptical. We didn't have radiation readings in familiar terms. Some wondered if the Soviets were minimizing the danger to pacify our fears. Patricia made attempts to contact various American officials to confirm what the Soviets had told us. Although everyone we spoke with was reassuring, my anxiety continued until I received a thorough check in Arizona after returning home. As it turned out, what the Soviets had told us in Moscow was confirmed by our own experts.

* * * * *

From Moscow we traveled by Finnish motorcoach toward Novgorod and Leningrad and stopped briefly in Kalinin to visit the home of Tchaikowsky. Two weeks previously, the great Russian pianist Vladamir Horowitz had made his historic return to his homeland and stopped here. Tchaikowsky's piano usually remains silent except for once every four years when the winner of the Tchaikowsky competition has the honor of playing it. Making a special exception, the Soviets were honored to let Horowitz play it during his visit.

We spent the night in Novgorod, one of Russia's oldest cities, dating back to at least 859 AD. Here we learned of the Russians' legendary reputation for endurance and ability to survive under adverse circumstances. During the 900-day siege of Leningrad and Novgorod by the Nazis in World War II, the population of Novgorod shrank from 4000 prior to the bombardment, to 38 after. Yet, despite overwhelming casualties, they didn't surrender.

In Leningrad, we again had the opportunity of meeting with the local Peace Committee. We broke into small discussion groups and had the chance to communicate with Soviet citizens from various walks of life, including lawyers, doctors, nurses, and educators. There was a large workshop devoted to discussing foreign relations. Here we

Young Muscovite

Through the eyes of a child we will see the path to peace.

had another opportunity to bring up questions about Afghanistan, religious repression, human rights, nuclear disarmament, and American fears of Soviet world domination. Again, I found the answers intellectually satisfying and not just regurgitated party rhetoric. It became clear to me that there is more than one way of viewing these complex issues despite our tendency of wanting to reduce everything to black and white.

During the remainder of our time in Leningrad, we continued to meet ordinary people on the street. The more we talked, the more it became clear that we were all citizens of the same planet with common interests outnumbering our differences. True, we grew up in different cultures, spoke different languages, lived under different governments and political doctrines, but underneath it all we faced the same human predicament. We are born, we play, we work, we dream, we raise our children, we make friends. We learn, we love, we grow in wisdom and understanding, we enjoy the treasures of the world, and eventually (for now anyway) we die. We all share the same desire to live peacefully together, to leave our children a world free of the threat of war with one nation against another.

In the old style of thinking, based on duality, peace becomes a tremendously complex and improbable event. How can so vastly different cultures with incompatible ideologies reconcile with each other and prosper from their coexistence on the planet together? In actuality, when we stand before each other face to face the process becomes easy. Then we are able to transcend our differences and find that what we share in common is more powerful than the differences that separate us. The key, however, is face-to-face personal contact and thinking of each other as equals.

Dennis L. Haughton, M.D.

Do you know what the most profound revelation that came to me as I shared and interacted with the Soviets? Beneath the misunderstanding, the suspicions, the mistrust, the fears, and the denunciations that have separated us, there is a bridge of love that connects us that is more powerful than all the nuclear missiles we hide behind. It is a bond of understanding that transcends the forever of time and space. Beyond words, we can feel it within our hearts and know it in our souls.

And then, before I knew it, it was time to pack up the suitcases again, say goodbye to my new Soviet friends, and once again head back across the border to Finland. It was somewhat of a cultural shock to leave the simplicity and relative austerity of Soviet life and once again be surrounded by the affluence and infinite variety of Western culture. I can remember walking through one of the new shopping malls in Helsinki and thinking that for the average Soviet citizen, opulence and diversity like this were unknown, although for me it was routine. I thought of my friend Anatoli in Moscow, who, although having little material wealth felt fulfilled simply because he had God in his heart. How often, I wondered, do we miss seeing the forest because of all the trees?

Starving for news about Chernobyl, our group nearly bought every issue of *Time* and *Newsweek* available at the airport while awaiting our departure for the States. Now for the first time since Kiev we could read about all the details of the disaster that we had lived through yet were kept uninformed about.

We were scanned for radiation at the Finish border, so I was a little surprised when no one met us in Los Angeles or New York to check us upon our return to the States. In both airports, customs just waived us through; it was as if they didn't even want to touch our bags. This arms-length treatment continued when we arrived in our home towns. Many in our group found it very difficult to persuade anyone to check them for radiation.

I was one of the luckier ones, however. Perhaps due to my connections in the medical field, I had no trouble making the arrangements to be checked. The day after my return, I phoned the nuclear medicine department at the medical center in Phoenix where I work. My request was unusual and I think I caused a panic when I asked if I could come in to be scanned. When I arrived, a security guard met my car in the parking lot and wouldn't let me out until Jim, the head technician, arrived with his Geiger counter. Only after he was reassured by a quick scan while I sat in my car was I allowed to get out. I then had to put on a surgery gown, cap, and boots so I wouldn't shed my radiation inside the building. I asked him to scan my luggage as I was most concerned that my film might have been destroyed. The racket from his instrument as he held the probe next to my camera bag was not very reassuring. When I asked him to scan the film canisters inside he refused, saying that the readings were so high that he was positive that the film would be hopelessly fogged. With that, my heart sank ten feet into the ground.

Following gloomily behind, I was led into the X-ray waiting room and told to sit in the corner with my arms crossed away from everyone else. (I wondered if this was how lepers feel.) Finally, when he had carried out all necessary precautions, I was taken back to the scanning room and positioned under one of the larger gamma cameras. Since a large part of the fallout from Chernobyl is in the form of Iodine 131, which is concentrated by the thyroid gland, a measurement from there would give a rough approximation of the total body dose.

As the time ticked away, and I anxiously awaited the outcome, and hence my fate, my optimism dwindled as I heard him making comments like, "Wow, look at this thing light up!" as he watched the monitors in back of me. Lying there patiently, I closed my eyes and watched an abbreviated version of my future flash in front of me. Finally it was over, and he announced that actually, I wasn't registering much higher than ordinary background radiation. We then both breathed a sigh of relief (I more than him), I took off my surgical attire, and canceled my plans for funeral arrangements.

I was now reasonably assured that, as the Soviets had told me in Moscow, I would probably not suffer any major health consequences from my fallout exposure. My biggest concern now was that my film was destroyed and I wouldn't be able to share my photographs of the Soviet Union with anyone. (At this point I hadn't even thought of writing a book.) So, I got into my car and drove to the Arizona Radiation Regulatory Agency to have my luggage checked.

I am deeply grateful to the people at ARRA for the professional and speedy way in which they tested my belongings. When I got there I guessed from the excitement present at their facility that testing fallout victims was a rare event. As four or five scientists took apart and probed my luggage, a photographer recorded the event on film.

(Here it is necessary to digress a little and summarize a few facts about radiation. Natural environmental sources of radiation such as the sun, cosmic rays, radioisotopes occurring in rocks, and radon gas produce an annual average background dose of between 200-300 millirems per year. Artificial sources of radiation, primarily from the medical use of X-rays add another 50 millirems per year on the average. This totals an average dose from all sources of 300 millirems per year, or about 0.04 millirems per hour. Keeping these figures for average background radiation in mind will make any figures that follow more meaningful.)

When all the tests were done, they had both good news and bad news for me. The good news was that my film probably was safe. Although there were relatively high readings on the outside of my camera bag, the inside where my film was stored was clean. Apparently, the external radiation was almost all beta radiation (consisting of free electrons) which can be stopped by a piece of thick cardboard. The nylon material of my bag was more than adequate to shield my film within. Had it been gamma radiation (a burst of pure energy, much like X-rays), nothing short of thick concrete or lead shielding would have stopped it. Since concrete camera bags are rather impractical, I was glad at least that we were dealing with beta radiation.

The bad news was that some of my things were too "hot" to take home. My brand new pair of Nike running shoes (the same ones coveted by my Ukranian friend Mikhail) had the highest readings of 3 millirems per hour. A new corduroy sport coat, my camera bag, several shirts, the earphones to my portable stereo, and a few other things also were highly contaminated and had to be left behind. They performed a swipe test to see if the radiation could be removed (basically this involves wiping a sticky piece of paper across the item and testing the paper for any transferred radioactivity), but unfortunately none come off. Then they confiscated the items since they knew of no practical method to decontaminate things once they were tainted.

One of the scientists let me take his personal Geiger counter home with me to test each item in my suitcases. Meticulously, I scanned everything that I had brought back from the Soviet Union and found some surprises. Several things that were never outside in Kiev were contaminated. I found that the distribution of radioactivity was not uniform. For instance, several shirts that overall gave low readings had isolated spots of intense radiation the size of a quarter where a single fallout particle had lodged. I wondered what the long term effect would be of having that spot in prolonged contact with my skin. I decided not to perform the experiment, however, and threw them away.

The other surprise was that after several trips through my washer, the clothing was just as hot as it was before. I kept careful records of my observations. Several areas on my hair remained radioactive for about a month, despite daily washing. Fortunately, the majority of radioactivity was due to the isotope Iodine 131, which has a half-life of around eight days. Roughly speaking, this means that after eight days only half the original amount remains, after eight more days, only a quarter, after another eight, only an eighth, and so on.

One of the radioisotopes from Chernobyl found in my luggage however, was Cesium 137 which has a half-life of 30 years, meaning that it stays around for a long, long time. Despite reassurances from nuclear experts I wonder about the long term consequences of a Cesium particle if it lodged, for instance, in my lung. Perhaps insignificant to my whole body, what would the effect be on cells in the immediate vicinity of that particle?

Certain conclusions seem obvious. Radioactive fallout is colorless, tasteless, odorless, and invisible. Once it escapes into our environment, we have to live with it until it decays, a process that can take hundreds or thousands of years or longer. It can't be dislodged, removed, or neutralized easily. It poisons living creatures. In short, we don't want anymore of it unleashed on our planet and into our environment—either accidentally like Chernobyl or intentionally with nuclear weapons.

Several other members of my group also had fairly complete examinations when they got home. A few had whole body scans at local nuclear power plants which detected low levels of radiation. So far, the highest readings to my knowledge were on Wendy's shoes since she spent a great deal of time outside in Kiev videotaping our activities. Her readings of 100 millirems per hour were more than 100 times the official level disclosed by the Soviet authorities for the Kiev region and several thousands times background radiation norms.

Many, however, had difficulty persuading local authorities to check them at all. Advice from doctors ranged from "Go home and take two aspirin" to "Go home and make out your will." Erroneous advice such as telling people to put their irradiated belongings in a plastic bag for a month or to just wash things well before wearing them again was common. One lady from Southern California who was checked was told that the levels on her clothing were not dangerous but they confiscated 95% of her things anyway.

Even before any specific readings from the Kiev area were made public, people in our group were reassured that they didn't need testing because their exposure was safe and they had nothing to worry about. Many authorities downplayed the whole incident, often without specific data upon which to base their assurances. In a medical journal several months later, an anonymous source from one nuclear power plant explained that the reluctance to check people was based more on liability concerns, rather than the lack of concern for the victims. After having been told that everything was OK, if a person were to later develop cancer, a costly lawsuit might ensue. Even if the doses of radiation we received were small, as it turned out, there was no justification for the experts to offer their reassurances without even testing people or having reliable information about exposure in the fallout zone. As I later would learn, this was not a unique occurrence. Government and nuclear industry spokespersons the world over have a long history of understating, downplaying, and in some cases frankly lying to the public about the potential dangers of nuclear power.

Many of us became instant celebrities when we came home as the media was hungry for first-hand information from people who had actually been there. I was written about in five or six articles in local newspapers and in two national medical journals. My most interesting adventure was being interviewed on a local radio talk show. The host couldn't believe why I was not totally enraged at the Soviets for allowing me to stay in a fallout zone. I really could hardly blame anyone for letting the accident happen and then for not giving a few Americans in Kiev a higher priority for evacuation than the 100,000 people in the immediate area around the reactor. While driving home after the interview, I listened to the reactions of subsequent callers. Several of the more opinionated callers as well as the talk show host called me a "Commie" for not being more anti-Soviet. I get a chuckle from this when I think about it now.

It has been a year since my return and I remain in excellent health. My hair hasn't fallen out and I don't glow in the dark, although many people still ask if I do. Except for one woman that I know of from Hawaii, the rest of the group are well. I don't expect any long term health consequences to result from my venture in the Chernobyl rain. From all that I know, my odds of getting cancer are extremely remote. Even though I am several thousand times more likely to develop a tumor than winning the Arizona lottery, I'm not worried much. It's the people in the immediate vicinity of Chernobyl primarily that will pay the price for the mistakes that were made.

I have given several public slide shows of my experience and spoken to hundreds of Americans about what I have learned. I keep a set of pictures from my trip hanging in my exam rooms in my office, and am surprised at the number of patients expressing an interest. Most people I talk with seem open to a new understanding of relations between Soviets and Americans. I am encouraged about the prospects of real peace on our planet arising from this new awareness.

I have kept in contact with other ICU members with whom I traveled. I have become much better informed about the Soviet Union and nuclear energy and read everything I can find. I discovered a number of excellent documentaries about Soviet life appearing on educational TV that were refreshingly free of the usual anti-Soviet prejudices. I was fascinated by the U.S. - Soviet talks in 1986 that built up to the summit meeting in Iceland. Real progress toward freeing our planet from threat of nuclear war now seems possible.

I know there must have been a purpose for 60 Americans to be caught up with Soviets in the Chernobyl disaster over a year ago. I have grown considerably in wisdom since then and I have learned that I am not powerless to change the present reality in which I live. I have made a commitment to support the efforts of other citizen diplomacy groups in bringing about planetary healing. As of this writing, I will soon be leaving on another trip to the Soviet Union. This time I will be traveling with a group of psychologists and other professionals, and we will be holding workshops with the Soviet people exploring our feelings about each other in our attempt to improve our communication and understanding that we are citizens of the same planet. We now stand at a crossroad in human evolution. We can continue on as we have been, a people divided against each other, and muck up things even further as we pursue our petty nationalistic rivalries, or we can begin forming the visions of tomorrow that will lead our children into an age of unprecedented peace and planetary understanding.

CHAPTER SIX

The aftermath: from out of the ashes

Over a year has lapsed since that fateful night in April when uncontrolled and runaway nuclear forces ripped apart the core of reactor four and spewed its toxic innards onto a stunned world. The radioactive cloud has long since dispersed, political fallout has died away, and the world goes on with the process of healing its wounds. Have we learned anything from the experience? Can a nuclear disaster of such magnitude happen again? And what relevance did the incident have to mankind's huge arsenal of nuclear weapons? Unlike power reactors, bombs have no containment structure and, by design, eject all of their poisons into the winds—not to mention the unhealthy effects they have on the poor souls standing directly under them at ground zero when they detonate.

Despite all our sophisticated scientific achievements, we still settle our disputes like barbarians by killing or threatening to kill each other. National strength is too often equated with numbers of soldiers, tanks, and megatons. Often we are so caught up in our creeds, nationalistic prides, causes, and personal dramas, that we forget what peace really means. Before we go any further, let's take a look for a moment at where we are now, how we got there, and how we can use the wisdom gained from our experience to move toward global health.

The explosion of reactor number four and its subsequent release of radioactive fallout, highly contaminated an area around Chernobyl of roughly 1000 square miles, covering the Ukraine and three other Soviet Republics. Despite massive cleanup efforts and the optimism of Soviet scientists, it is unlikely that the area can be safely inhabited or farmed for the next 10-30 years. Levels of radiation around the reactor itself were astronomically high. Attempting to minimize the danger and reassure a nervous population, Soviet officials originally admitted to figures of 20-30 millirems per hour. Later in Vienna, during the inquiry by the International Atomic Energy Agency (IAEA) in August, they reported levels in "excess" of 100 millirems per hour. In fact we now know that the levels were millions of times higher, in the order of hundreds of thousands of *rems* per hour. Radiation was so intense that they couldn't locate any instruments in the world that were capable of measuring it.

The lessons learned from the experience of evacuating the contaminated area has demonstrated woeful inadequacies in emergency plans throughout the rest of the nuclear world. 12 hours after the accident the first wave of evacuation took place and roughly 1000 families living within a mile of the plant were taken away by local transportation. The next day, about 36 hours after, 46,000 additional people were evacuated from the town of Pripyat and the area within a radius of six miles. Then a dangerous delay occurred due to confusion and poor decisions on the part of local Ukrainian authorities. It wasn't until nine days after the accident that the 30,000 inhabitants of the town of Chernobyl, as well as thousands more inhabitants within a radius of 19 miles from the plant, were evacuated. This delay caused serious radiation exposure that wouldn't have occurred had the evacuation been more timely.

In all, around 135,000 residents were evacuated from the 19 mile zone around the damaged reactor. People had to abandon most of their belongings and had to be fed, housed, and relocated. In addition there were over 17,000 head of cattle that had to be relocated, fed, and milked. Today, the towns of Pripyat, Chernobyl, and Lelev remain empty ghostowns. 60 miles south in Kiev, 250,000 children were sent away to distant camps or resorts for the summer. Had wind and rain carried the initial radioactive cloud south, the two and a half million residents of Kiev would have had to be evacuated also.

The spread of the cloud observed no national boundaries. Fallout occurred in virtually every country in Europe. Millions panicked not knowing how much longer the radioactive rain would continue. It covered Europe in a very haphazard pattern, causing dangerously high levels in areas hundreds of miles away. Its presence over people's heads caused an acute awareness of our mutual vulnerability and gave many Europeans a frightening preview of the possible effects of a small nuclear war.

And yet despite the damage caused, it could have been far worse. Due to a number of lucky factors, the most deadly part of the initial plume probably missed heavily populated areas. The winds carried the first discharge towards unpopulated swamps, just missing Pripyat. The fierce heat of the fire coupled with the still night air allowed the deadly smoke to rise straight up where it was carried northwest and away from the immediate area. There was no rain, allowing short lived radioisotopes to decay harmlessly in the atmosphere the longer they stayed aloft. The prevailing winds for the most part directed the largest portion of the cloud away from heavily populated Kiev. Since the accident occurred at night, most people were inside and had some protection by their homes. Rather than thousands of workers there were only hundreds working at the plant during the night shift. In a worst case scenario, thousands of immediate deaths could have occurred, not just 31.

One expert estimated that the fallout from Chernobyl was equal to the fallout from all previous atmospheric weapons tests. In Finland, more that 1000 miles distant, radiation levels were ten times normal. In areas of Poland, levels reached 500 times normal. A severe shortage of honey was predicted there because of the millions of bees killed by the cloud. In Paris, fallout in early May was as high as it was during the peak of atmospheric bomb testing in 1963. Milk was banned 800 miles away in Sweden. Reindeer in Lapland are still unsafe to eat today. Fish in some small lakes in Finland remain inedible. The list goes on and on.

The response by many governments was chaotic, disorganized, and often dishonest. The East Germans simply announced that there was no danger from the fallout, yet gave no details about radiation levels. The French concealed levels of contamination 400 times normal from the population. The British government bungled the dispersal of timely information causing public confusion and alarm to rise to high levels. Evaluation of civil nuclear disaster plans were found to be dangerously inadequate. A recent issue of *Newsweek* magazine reported that in Britain today, many communities are considering installing their own radiation monitors because they can't trust their governments to tell the truth about future leaks. It further reported that in Greece, a group of doctors, scientists, and lawyers plan on measuring radiation levels in food themselves, not trusting public officials to inform citizens about contaminated produce.

Massive cleanup and decontamination efforts were undertaken by Soviet experts to prevent the further release of radioactive poisons from the damaged reactor and to restore operation of the other three reactors at the power plant. Reactor four has now been permanently entombed in concrete and steel where it will remain forever, despite optimism of restoring operation some day. Experience with the Three Mile Island accident in Pennsylvania has shown that decontamination is exceedingly difficult and prohibitively expensive. Cleanup at TMI, originally estimated to cost $200 million over two years, has so far taken seven years and $1,000 million, yet the reactor may still remain in mothballs indefinitely.

The economic costs have been heavy. Although Soviet officials estimated a total cost of $2.8 billion, Western experts predict a much higher figure, between $7 and $14 billion. Eastern Europe alone lost an estimated $975 million in exports. On a more positive note, Soviet citizens coming to the aid of their comrades raised nearly $450 million to help the victims of the accident who were injured or had to be relocated.

The economic penalties, however, are small in comparison to the projected long term health effects. Radiation kills in a number of ways. Massive doses can kill in a matter of minutes to hours by frying the brain and central nervous system. Slightly lower doses create acute radiation sickness, whose victims usually die within a week or two, primarily due to damage to the gastrointestinal system. Those who recover from acute radiation sickness, or have received less toxic doses, die in a month or two after exposure due to bone marrow failure, causing the victims to lose their ability to make red blood cells, white blood cells to fight infection, and platelets to control bleeding. So far these first three waves of death have claimed only 31 lives, primarily workers and firefighters who received massive doses in the first few hours after the initial explosion.

Present treatment measures for acute radiation exposure are not very effective. Despite heroric efforts by the American doctor Robert Gale and his team of specialists flown in to provide emergency aid, only five of the thirteen bone marrow transplant recipients survived. The procedure is not only complex but expensive; each transplant costing up to $100,000. Later analysis of the transplant cases has raised serious doubts about the value of this procedure under this set of circumstances.

In all, there were a little over 200 acute radiation victims. The majority survived, but even this small number severely taxed the ability of the Soviet medical system. Intensive care facilities had to be improvised and Soviet officials readily admitted there was a shortage of doctors trained in this field. Had conditions at the time been less favorable and there had been thousands of acutely irradiated victims, the system would have been completely overwhelmed. In such a situation, the majority of the victims would simply not have received the intensive critical care necessary to survive.

The Soviet's inadequacies or inferior equipment and facilities are not to be blamed, for many have estimated that the American medical system would also be severely strained by an accident of similar magnitude. Keep in mind, we are still dealing only with a small scale localized nuclear disaster. In a nuclear war we would be dealing with millions of victims, not hundreds. Forgetting for a moment the millions killed instantly by the blast and the ensuing raging firestorms, the vast majority of survivors with acute radiation sickness and excessive burns would receive virtually no advanced care except for a few rudimentary first aid measures. Even assuming that all the hospitals, intensive care and burn units, and medical personnel miraculously survived the blast in a major U.S. city, there still could be well over 100,000 burn and radiation victims needing the services of only several hundred critical care beds.

The majority of deaths from Chernobyl will occur in a fourth and fifth wave of long term malignancies. From two to 25 years later we will see a wave of leukemias, primarily in children and young people. Between 10 and 40 years later we will see a wave of solid cancers including multiple organ systems. This means that in the year 2030, people will still be dying from the effects of Chernobyl. Even today, bomb survivors from Hiroshima and Nagasaki are still contracting fatal cancers from radiation received more than 40 years ago.

There is no agreement on the exact number of long term deaths that will occur. The Soviets' report to the IAEA in August estimated that 6,300 people ultimately would die of cancers. I have seen some estimates as ludicrously low as 171 to as high as 280,000. Figures between 30,000 and 50,000 are more likely. A number of factors make exact predictions difficult. No one knows for sure how much of reactor four's core went up in smoke, how much of it landed, or even where all of it landed, since distribution was uneven. Another problem is that the calculated cancer rate of Hiroshima victims, used in projecting fatalities for this disaster, was originally based on higher radiation doses than they may have actually received. If lower doses actually caused the same number of cancers in Japanese victims, then the Chernobyl fatalities will be much higher.

Although the majority of casualties will originate in the most heavily irradiated portions of western U.S.S.R., sizable numbers will occur in areas far removed from Chernobyl. The Poles estimated up to 500 deaths, the Swedish up to 150, and even official British estimates of 45 deaths were predicted. Another 3000 deaths have been estimated in western European countries. It is even conceivable that a few deaths could occur in the U.S. which also detected Chernobyl fallout. No one really seems to know what the long term genetic effects will be. One source suggested that it would take 50 generations before the full impact of genetic damage from Chernobyl would be known. Keep in mind these casualty figures apply to an isolated reactor accident under favorable conditions. In a more densely populated area with less favorable weather conditions, the casualties could be higher by several orders of magnitude. In the aftermath of a nuclear war, the long term casualties from malignancies alone would be astronomical.

Well, so much for Chernobyl for the time being. What about the other 375 or so nuclear reactors scattered around the planet? The U.S. leads the world with 93 operational and another 26 under construction. The Soviet Union comes in second with 51 operational, 34 under construction and another 39 in planning stages. Although France is third with 43 in current use and another 19 under construction, it currently derives 65% of its total power from nuclear energy compared with 15% and 10% for the U.S. and Soviet Union respectively. Britain and Japan are fourth and fifth and so on and so forth. Could any of these reactors blow up also?

If we listen to the confident pronouncements of nuclear industry spokespersons and government officials, we are told never in 100,000 years. Prior to the Three Mile Island (TMI) accident in 1979 near Harrisburg, Pennsylvania, experts assured us that the chances of a meltdown were too small to calculate. Their favorite phrase was "It could never happen." Afterwards, shaken somewhat by the partial meltdown in TMI's unit two, nuclear industry spokesmen then vowed "It could never happen *again*." Now, after Chernobyl, the U.S. nuclear industry assures us that although such an accident could occur in the Soviet Union, "It can never happen *here* again."

If anything, Chernobyl has reminded us that at the core of every nuclear reactor lies a brew of deadly poisons that retain their virulence for thousands of years and can not be captured once they leak into the environment. Furthermore, we still have no safe burying ground on the planet in which they can decay in peace for centuries. Shortly after the Chernobyl incident government and industry spokesmen were quick to blame the accident on shortcomings of Soviet nuclear technology. They asserted that Chernobyl was poorly designed, had no containment structure, and had inadequate emergency back up systems. American reactors, on the other hand, they said, all had containment structures, were much better designed, had a redundant number of emergency systems, and were virtually immune to an accident like Chernobyl.

In fact, later analysis of the Chernobyl reactor blueprints revealed that it was much better designed than originally thought. Far from having none, reactor four had two containment structures: one was built to withstand pressures of up to 27 pounds per square inch while the other up to 57 psi. American reactors generally have containment vessels capable of withstanding pressures between 45 and 60 psi. However, there are one or two U.S. reactors rated to only 12 to 15 psi and five reactors with no containment domes at all. These five are run by the Department of Energy and are used to produce plutonium for nuclear weapons. One of these, the reactor in Hanford, Washington, has a graphite moderated design, similar to that at the Chernobyl complex.

At least one type of U.S. reactor containment structure is considered by many experts to be unsafe. The Mark I boiling water type reactor built by General Electric has a containment structure that has up to a 90% possibility of cracking open in a major accident and venting enormous amounts of radioactivity into the environment. That estimate comes not from an anti-nuclear activist but from a chief reactor regulator for the Nuclear Regulatory Commission (NRC). There are now 24 Mark I type plants licensed or under construction scattered across 14 different states. Despite this shortcoming, the NRC continues to license reactors with this unsafe design. Last year, the 1067 megawatt Hope Creek Plant in Salem, New Jersey was granted final authority to begin operation. It has a Mark I containment structure.

In the final analysis, the Soviet accident occurred not so much because of technological shortcomings, but due to human folly which respects no national boundaries. If the operators of unit four had omitted just one of their fatal six mistakes, the accident would never have occurred. One observer likened the Chernobyl operators' experiment that night to a pilot in mid-flight turning off the engines and opening the doors of the plane to see what would happen.

Some have charged that nuclear regulatory agencies often develop an unhealthy symbiosis with the government and nuclear industry they were created to supervise. Rather than protecting the public welfare by ensuring safe reactor design and operations, they are often seen to promote and protect the industry's economic investments. The nuclear industry throughout the world, irregardless of politics, has been engaged in an effort to cover up the risks of nuclear power.

In 1956, the Atomic Energy Commission (AEC - predecessor to the NRC) conducted a study estimating that a meltdown in an average-sized reactor would cause 3,400 deaths, 43,000 injuries, and $7 billion in property damage. This study, WASH-740, was updated in 1964. The revised report predicted 45,000 deaths, 100,000 injuries, $17 billion in property damage (in 1965 dollars), and would contaminate an area the size of Pennsylvania. The results were even more unpalatable to AEC. They had commissioned the study in the first place to convince the public and insurance companies of the virtues and safety of atomic energy. The results were so horrifying that they were suppressed until 1973, when the threat of a lawsuit by a Chicago lawyer under the Freedom of Information Act forced the AEC to reveal its findings. Similar studies in other countries also have been suppressed. (As an aside, in case you are living next to a reactor, read the fine print in your insurance policies. Very few insurance companies are foolish enough to protect you from the risks of a nuclear accident.)

What about other accidents in the past? Chernobyl was not the first major Soviet accident. In late 1957 or early 1958 several hundred square miles around the plutonium producing complex at Kyshtym in the Southern Urals were contaminated with lethal levels of radioactivity. The exact cause is still not known as Soviet officials have not yet acknowledged that it even occurred. It is not known how many casualties there were, but it is likely that several thousand people were evacuated. Widely accepted by western experts as the largest nuclear accident before Chernobyl, the U.S. government did not publicize it until 1977 when attention was drawn to it in an article appearing in a British publication, *The New Scientist.*

At the Windscale reactor in Britain, famous for setting the record for the most radioactive discharges into the environment, there have been over 300 accidents. Nearly a quarter ton of plutonium has been discharged into the Irish sea and will remain intensely radioactive for 250,000 years. Animals grazing near the plant have accumulated levels of radiation hundreds of times normal. The worst accident occurred in October 1957, when a fire blazed out of control for 42 hours. Using huge volumes of water to extinguish the burning uranium, fuel cladding, and graphite core, the firefighters risked a larger explosion that could have blown the reactor apart. Despite the precarious situation and the escape of considerable radioactivity, the British people were not informed until the fire was nearly out. No one was evacuated from the surrounding area and complete details of the incident were never released to the public. Today, like reactor four at Chernobyl, the Windscale reactor lies entombed in concrete, never to be used again.

American reactors also have had their share of accidents and near catastrophes. Between 1969 and 1979 there were 169 incidents that could have resulted in a meltdown according to one government study. At the Brown's Ferry reactor in Alabama, a fire that raged for seven hours was started when an electrician lit a candle to detect a draft beneath the control room. All five emergency cooling systems were disabled and disaster was narrowly averted.

At another reactor, a basketball was used to plug up a pipe which later spilled 14,000 gallons of radioactive water. A 3000 gallon radioactive waste tank at another plant was connected inadvertently to a drinking fountain. Safety equipment installed backwards or upside down, an entire welding rig left inside a reactor blocking water flow, and the list of horror stories goes on and on.

But by far the worst and most famous American accident occurred at Three Mile Island in 1979. As luck would have it, I lived about six miles away from this reactor while I was attending medical school in nearby Hershey, Pennsylvania. As in the Chernobyl incident, the seriousness of the TMI accident was withheld from the public for three days. There were some striking parallels to the movie, *The China Syndrome,* which had been released just two weeks prior to the accident. Reassuring statements from the plant's public relation's spokesmen continuously downplayed the critical nature of the problem, while, inside the plant, no one had any idea of how to control the situation. In the end, it was learned that the reactor came within an hour of a full meltdown and it was only through sheer luck that it had been avoided. Despite recommendations by NRC at the time to begin evacuating the surrounding area before it was too late, the local commissioners decided against it. Today, the reactor remains dormant despite years of decontamination efforts costing nearly a billion dollars.

If anything, the Chernobyl disaster may force some honesty into the nuclear industry which for years has minimized the dangers of nuclear energy through secrecy and misinformation. Gone is the enthusiasm and optimism of the 50's that envisioned nuclear power as the cheap answer to mankind's energy needs. Cost overruns have made reactor construction in many cases unprofitable. Plants budgeted to cost $450 million have ultimately cost $4.4 billion. Between 1980 and 1982 40 plants, either under construction or in planning, were scrapped in the United States. There have been no orders for new nuclear reactors in the United States since 1978.

As to whether or not we can expect Chernobyl type disasters in the future, I'll let the experts speak for themselves. In the book *Chernobyl: the End of the Nuclear Dream*, by Hawkes, et. al., the authors state that "senior figures at both the U.S. Nuclear Regulatory Commission and the IAEA privately believe that we can expect a major accident every decade." The NRC has reluctantly admitted that the chances of a major accident occurring in the United States in the next 20 years is around 45 percent. These are hardly acceptable odds.

I want neither to completely condemn nor eliminate the peaceful use of nuclear power on our planet. It is possible that some day we may design a safe reactor, but that day is not yet here. In the meantime, I think we need to take an honest and hard look at the Pandora's box we opened when we decided to use Einstein's $E=mc^2$ before we had the wisdom to use it safely for the benefit for all humanity.

However, it isn't the peaceful use of nuclear power that poses the biggest threat to our planet. However horrible the consequences of Chernobyl, they are minuscule compared to even the limited use of nuclear weapons against each other. After all, nuclear reactors which are designed by the best minds around us to safely contain their lethal biproducts occasionally fail. Nuclear weapons, on the other hand, have been purposely designed by our best scientists to unleash upon us the totality of the atom's destructive power.

In a television address on May 14, 1986, Mikhail Gorbachev told his viewers: "The accident at Chernobyl has once again illustrated what a great abyss would open up if nuclear war should befall mankind. Indeed, the stockpiled nuclear arsenals are fraught with thousands of catastrophes much more terrible than Chernobyl." I never have understood why this was so quickly dismissed as propaganda the next day by officials in the U.S. government and the American media. After all, it sounded to me like he was just stating common sense.

In one of the greatest ironies of this planet's 4.5 billion year history, the dominant species (who for some reason consider themselves the most intelligent) have squandered vast amounts of the planet's resources in building weapons of mass destruction that threatens the basic life support system of the planet itself. Fueled by massive inferiority complexes and money extorted from the planet's inhabitants, the two competing superpowers have amassed a gigantic pile of more than 50,000 bombs—roughly equal to 16 billion tons of TNT or about 4 tons of TNT per person if equally distributed.

As if this wasn't enough, both sides are constantly afraid that the other's pile is either bigger, better, or more accurate than their own. Periodically, paranoid warlords from each tribe stand up and give frightening speeches about the malicious nature of the other side, thus fanning exaggerated fantasies of each other and ensuring that the ignorant masses will keep paying for the whole ridiculous mess. Meanwhile, straight faced politicians do their best to convince everyone that the whole process is quite rational while hiding the fact that this mine-is-bigger-than-yours contest is quickly driving the world toward bankruptcy.

Despite having less than 11 percent of the world's population, the two superpowers account for 23 percent of the planet's armed forces, 60 percent of its military expenditures, 80 percent of the weapons research and development, and 97 percent of the nuclear warheads. Mortally threatened by each other's political creeds, they compete endlessly to convince other nations on the planet that their system is better than the other's. In order to win friends or to raise needed cash, they sell their advanced weaponry to less developed tribes that otherwise would be forced to kill each other with primitive sticks, stones, and surplus World War II armaments.

In order to pay for the whole thing, one superpower makes its citizens go without the luxuries of shiny new cars and fancy Japanese electronic toys while the other superpower borrows money from a generation of grandchildren who can't yet vote due to the fact that they haven't yet been born. Meanwhile, to raise cash to buy weapons, the less developed countries borrow money from anyone foolish enough to hand it over, while letting thousands of their own people die of starvation. Wonderfully caught in the middle of the whole complex operation are the owners and stockholders of the international weapons corporations—who, not too surprisingly, have bulging Swiss bank accounts.

The costs of creating, maintaining, and perpetuating the worlds' vast military machine are enormous. In 1986, we spent nearly $900 billion supporting our military habits. That's nearly two million dollars per minute, every minute of the year, or roughly 2.5 billion dollars a day. The only way we have been able to afford such extravagant weapons purchases is to rob precious resources from needed social programs and generate a huge pyramid of debt to pass on to future generations. Often with higher priorities for weapons than for people, we are willing to let millions starve, go homeless, or grow up ignorant so that we can have the newest model tanks, bombers, and rockets.

According to Carl Sagan in a recent article, in *Parade* magazine, from the end of World War II until the time Mr. Reagan leaves office in January 1989, Americans will have spent nearly ten trillion dollars on its global confrontation with the Soviet Union. That's enough money to buy everything in the United States except for the land. Think of it. All the houses, bridges, TV sets, cars, planes, factories, highways, skyscrapers, boats, railroads, food, furniture, toys, underwear—*everything*! Think of what we could have done with even half that amount if spent for the benefit of us all. And today after squandering such a huge sum on national "defense," the Untied States is vulnerable to instant annihilation.

As our weapons grow in destructiveness, our wars have become more deadly. In the eighteenth and nineteenth centuries combined, roughly 13 million people were killed in wars. So far in the twentieth century, 99 million have perished. We now are confronted with weapons of destruction so powerful that their use against one another would likely destroy modern civilization as we know it in the span of a few hours. Yet because of this, for the first time in recorded history, we have been thrust on the threshold of a profound realization. Despite those in government and the military who persist in downplaying the consequences of nuclear war, people all over the planet are realizing that the use of these genocidal weapons against each other would mean racial suicide. Maybe we needed to come to the brink of annihilation to recognize our craziness. Ironically, perhaps nuclear weapons have given us the awareness that we can no longer afford to settle our international disputes by force.

Now that we have created this military monster and recognized its potential horror, we have yet to decide what to do with it. Some want to make it even bigger. Many military and political leaders still cling to the somewhat primitive concept of deterrence, which is based on mutual mistrust and fear. In simple terms, one superpower threatens the other by saying in effect, "If you attack us, we will attack you so viciously and kill so many of you that you would have to be suicidal to even think of shooting first." In this ultimate form of world terrorism, appropriately termed MAD (Mutual Assured Destruction), one government holds the entire population of the other hostage at nuclear missile point and vice versa.

Now the problem is that deterrence is based on the premise that the launch buttons for the missiles will always be in the control of rational human fingers. Unfortunately, there are those whose talk of striking first and "winning" a nuclear war by sacrificing hundreds of millions of people and civilization as we know it, borders on insanity to say the least—and what happens when somewhere in the innards of the vast computer networks that control the war machines, a fifty cent silicon chip fails and missiles are launched by accident? "Oh, it couldn't happen," they say. That's what we were told before Chernobyl and Three Mile Island.

Another approach for dealing with the nuclear threat which at first appears slightly more sane, is to build a perfect defense against the enemy's rockets. Here of course, I am referring to President Reagan's Star Wars or Strategic Defense Initiative (SDI). Envisioned and promoted as an inpenetrable peace shield over the United States, this miraculous marvel would destroy and prevent all of the enemy's missiles from ever reaching their targets. Such an impervious umbrella would, it was hoped, make offensive nuclear weapons impotent and obsolete. Close scientific scrutiny, however, has shot the umbrella full of holes. Two thirds of American physicists polled believed it was "improbable, or very unlikely." Far from making the world more secure, critics contend it would likely make nuclear war more probable because it could just as easily be used offensively in a first strike scenario.

We were privileged to have with us in Leningrad Dr. Robert Bowman, former director of the advanced space program development (the predecessor of SDI) in President Carter's administration. Dr. Bowman, who now travels extensively speaking out against Star Wars, calls SDI, "the greatest fraud perpetrated on the American people." Prohibitivly expensive, technically unfeasible, and impossible to test beforehand, its defenses could easily be defeated by countermeasures that are far cheaper. According to Dr.

Bowman, implementing Star Wars would require thousands of space laser battle stations each about the size of a football field. Their accuracy would have to be equivalent to mounting a machine gun on the Empire State Building in New York and hitting tennis balls bouncing around the courts in Wimbleton, England. One design would require three separate space mirrors per laser and its use would be equivalent to making a three cushion billiard shot over 50,000 miles.

Computer systems, vastly more sophisticated than anything currently available, would be needed to coordinate the intricate functioning of all its complex parts. The software alone would be more complex than any so far created by the human race. It would have to function perfectly the first time, but couldn't be tested and debugged beforehand under simulated battle conditions. Even if the system functioned perfectly and destroyed 90% of the incoming missiles, this would still permit the destruction of 1000 population centers. Dr. Bowman asserts that effective countermeasures could reduce Star Wars performance to roughly 10 percent. What is more, the system would provide no defense at all against low level bombers, surface hugging cruise missiles, or nuclear explosives smuggled into the country. Certainly, U.S. borders that allow hundreds of tons of drugs to leak through each year could not stop nuclear devices small enough to be carried by a single person.

Aside from the fact that Star Wars wouldn't work, it would cost about a trillion dollars or more to completely deploy, to say nothing of the costs of maintaining it. Considering that we have already spent trillions on an offensive nuclear weapon system that in all probability will never be used anyway, who wants to throw good money after bad? Think of all we could do with that $1,000 billion if we spent it on peaceful purposes. For that amount, we possibly could build a fleet of laser equipped weather control satellites which, by warming large areas of ocean or atmosphere, may be able to shift cloud systems and end the problems of draught on this planet. And even one billion dollars spent wisely promoting peace and friendship between the superpowers, could do more than all the weapons and missiles ever built to establish peace between our societies.

In our ignorance, we have built these war machines to protect us from each other, thinking that peace somehow depends on frightening off enemies that lurk only outside of our borders. We have given our power away to politicians, governments, national creeds, technology, and military might in hopes that somehow they could bring about a peaceful world for us. Not that the efforts by our statesmen to reverse a runaway arms race are not important. I commend the recent talks between Mr. Reagan and Mr. Gorbachev to bring about reductions of our nuclear arsenals.

But focusing on reducing or eliminating weapons is not enough. We get hung up on the minutia of counting megatons and the legal gobbledygook of treaties. Even now as we negotiate the elimination of medium and short range nuclear missiles in Europe, many European nations are anxious about giving up their nuclear security blanket because the underlying mistrust and suspicion have not yet been resolved. Treaties, no matter how comprehensive and well intentioned, will never build lasting peace. They can be brushed aside at the whim of future political leaders who have other designs.

Dennis L. Haughton, M.D.

Even if we did eliminate all nuclear weapons from the planet, how long do you think it would take to build new ones if international tensions erupted in conventional war? What meaning do treaties have between countries locked in a struggle to destroy each other? Our science is just in its infancy. As we probe even deeper into the secrets of the universe, we are bound to unlock energies that could make nuclear weapons seem like firecrackers. What we are doing is approaching the problem backwards: eliminating weapons in the hope that this will bring us closer to peace. If we first made peace with one another, weapons would no longer be needed and would quickly disappear.

A more fundamental approach is needed. Just as individual health is much more than the absence of disease, so is peace more than just the absence of war. Throughout recorded history, humanity has lived in a divided world. We have had brief periods without war, but we've never had lasting peace. Today, there are 20 different wars raging on the planet in which more than 1000 people are killed each year. The U.S. and the Soviet Union, while technically not at war with one another, are far from being at peace. Certainly countries aiming 10,000 nuclear missiles at each other and threatening each other with total annihilation can hardly be called friends.

Health on an individual basis requires the smooth functioning and harmonious integration of mind, body, feelings, and spirit. It is a state of wholeness in which the whole is greater than the sum of its parts. In this state, there is a zest for living, a joy in the moment to moment unfolding of life, a sense of fulfillment and tranquility, and an awareness of harmony with the universe. Disease is unable to gain a foothold. Internal struggles in which one aspect of the personality wars with another part have been resolved. Self-love replaces internal criticism and feelings of worthlessness. From this state of internal peace and wholeness peak experiences and heightened consciousness become common.

On a global scale, peace also implies wholeness. Rather than international rivalry, senseless antagonism between ideologies, and war, cooperation, mutual assistance, the tolerance and acceptance of differences, and global harmony can exist. National and political priorities become secondary to the needs of the planet and us all. Wars can be replaced by peaceful ties of friendship and brotherhood. Nationalistic pride and patriotism can make way for an emerging world consciousness. Starvation, ignorance, preventable illnesses, and unnecessary poverty can become history. From this state of planetary wholeness will arise human achievements and experiences that are beyond our present imagination. From this state of global interconnectedness a new age will be born as mankind truly lifts itself out of the age of ignorance and homo sapiens evolves into something more.

There are several reasons why we don't yet have peace on our planet. The most important of these is that we have an attitude problem. We simply believe in the inevitability of war and therefore it persists. For millennia we have been warring against one another and we continue out of habit. We don't yet realize that we can manifest whatever reality we choose on this planet. We can. We have seen ourselves as either the helpless victims of evil unconscious drives or somehow controlled by forces outside of ourselves. Nonsense. We created the present world reality by ourselves without anybody else's help. We have the complete freedom to do good or evil. Because of our belief in our powerlessness we have rendered ourselves impotent. We manifest what we believe so to change our reality we first must change our thoughts. In a brief moment of expanded awareness this could happen in an instant.

A second and related reason we don't have peace is that in order to rationalize the existence of evil in the world we convieniently blame it on somebody else. Glorifying our own immaculate virtues, we project onto our enemies all the vices and shortcomings we can't bear to recognize in ourself. American leaders, in condemning the Soviet government for the war in Afghanistan, forget their fiasco in Vietnam and our very questionable involvement in Nicaragua. Soviet leaders pointing their fingers at the "imperialists" in the U.S. conviently forget about Afghanistan. We think if it weren't for "them" the world would be better off. (Here you can substitute the names of anyone you have ever condemned, hated, judged, or felt better than).

Criminals of all types exist in all nations, in all races, under any creed, and with few exceptions, wherever humanity can be found. The holier-than-thou finger waving of the past and present will make way for forgiving tolerance as our consciousness expands. As we become more enlightened as individuals, we will recognize more and more that often what comes out of the mouths of politicians or others intent on blaming something on someone else is, quite simply, hot air.

The final reason we do not have peace is straightforward: we have not yet decided to do it. Imagine for a moment what would happen if Mr. Reagan and Mr. Gorbachev left their stuffy suits and advisers behind and just went off together on a three-day fishing trip in the mountains. In that relaxed atmosphere they learned to geniunely communicate and trust each other as human beings. From that state of emotional and spiritual connectedness, they pledged to lead their nations to that same state of trust from which peace can be automatic. I believe that in their past encounters, they have gotten close to this awareness but stumbled over how to solve all the remaining differences. What would happen if, rather than promoting misunderstanding and fear, they ordered the propaganda presses of both countries to promote peace and good will between Soviets and Americans? If we can program national consciousness to create hate and war, we can just as easily program it to create friendship.

The future we create will depend on the decisions we make today. Clearly, fundamental changes in how we relate to each other are called for. Perhaps from the ashes of Chernobyl will arise a new understanding of our interconnectedness that will help guide us in planning the 21st century. The choices seem obvious: continue war and perish as a race, persist in the present arms race and go bankrupt, or turn our weapons into plowshares and create a new earth.

CHAPTER SEVEN

To our children's children

There is a saying in Russia that I heard while visiting one of the public schools in Moscow:

> "Those who think one year ahead grow grain.
> Those who think ten years ahead grow trees.
> Those who think a hundred years ahead raise children."

Are we looking that far ahead? The dreams and visions we have today will create the world in which our children and their children who follow will live. The future is limited only by our vision of who we are and what we can become. What would happen if the parents of the world sat down together tomorrow and designed the future world that our grandchildren would inherit in 50 years? If, rather than conducting our planetary affairs based on short-term nationalistic goals, we based our decisions on the love we feel for our children in our hearts, would we allow war to exist? Would we want them to live on a polluted Earth? Think how different our world would become if we were designing it as a precious gift for our grandchildren. Imagine the wonderful world we could create together if only we bestowed that same type of love on ourselves. *We can.*

There is no doubt in my mind that the Soviets I met, as do Americans, love their children deeply. They consider them a national treasure. Everywhere I went and watched, parents and other adults were patient, firm, caring, and loving to the children. There almost seemed to be an element of collective parenting. Unattended infants in their carriages in front of GUMS, the largest department store in Moscow, suggested a level of group trust I hadn't encountered before. Somehow I couldn't imagine that happening in New York.

Look at the kids on the pages of this chapter. Do you know what they all have in common? They all look normal, don't they? Yet all of them are radioactive. As they innocently play with their balloons and splash in the fountains, an invisible and silent rain of deadly radioisotopes from Chernobyl covers them. These little ones are the most vulnerable to the effects of radiation. It gets on their clothes, settles in their hair, enters their lungs. It doesn't wash off easily, remember? They received a much higher dose than I did because I only was in Kiev for the first day of fallout. Some of them may die in the years to come because of what happened in those few days of April and May of 1986.

Just because these are Soviet children belonging to someone else, are they any less deserving of our love and good will than our own children? They are our kids also, for they belong to the future of all of us. Along with our own children, they will be the men and women who inherit our legacy. Let us not leave them a world poisoned by the madness of a nuclear war, a world that is thousands of times more radioactive than the one we have made them live in now.

All the children pictured here are standing in a radioactive rain of fallout from Chernobyl. These girls were singing a Ukrainian nursery rhyme as they passed.

Children of Kiev - 1 May, 1986

The red flags that they carry are the same as I had outside only during the parade. It was one of the hottest items I brought back home with me.

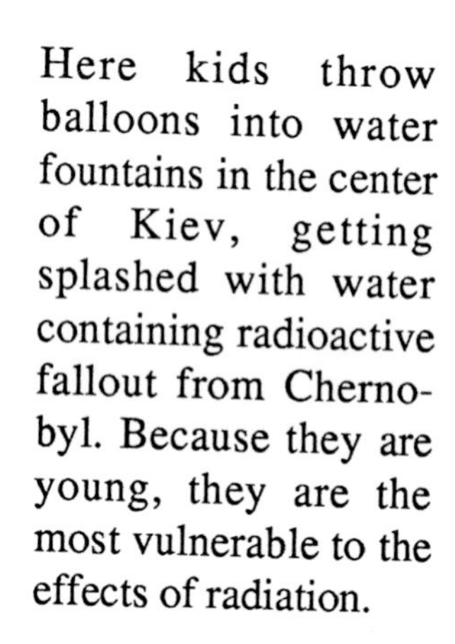

Here kids throw balloons into water fountains in the center of Kiev, getting splashed with water containing radioactive fallout from Chernobyl. Because they are young, they are the most vulnerable to the effects of radiation.

Long-lived radio-isotopes trapped in their bodies—like microscopic time bombs—could erupt forty years later and produce fatal malignancies.

For what we spend in one minute on the arms race, we could prevent the deaths of four million children a year.

Instead of fear and mistrust we can teach our children brotherhood and co-operation. Instead of training soldiers, we can raise peacemakers.

In the eyes of a child, the world is a vast playground for endless advertures.

Children, in their innocence, are closer to knowing how to make peace than we are—their supposed teachers.

The present global threat of nuclear extinction looming over our heads causes our—and undoubtedly our children's—vision of the future to be very shortsighted. The psychological impact of this menace has to be profound. If we can't count on being alive in twenty years then what difference does it make how we live our lives now? Live for the moment! Forget about plans spanning generations. Who cares about our heirs in 2087?

This psychological numbing effect has created a widespread feeling of powerlessness as people no longer believe they have control over their own destiny. Political leaders pessimistically bemoan the inevitability of war—as though they have no more control over their actions than teenage boys who, despite their belief that it will lead to insanity, can't control their compulsion to masturbate. As we learn how to empower ourselves, we shall discover that we never were powerless in the first place, we only believed that we were.

If we continue to believe that we must live out the same old script in which we all have played roles for the last few thousand years, the future could be very bleak indeed. Will nations continue to play global terrorism and keep "peace" only by the threat of annihilating each other? With 50 billion people living here in 100 years, will millions of kids die of starvation daily because military power is still more important than human needs?

What of the monstrous debts we are passing along to our grandchildren to finance a war machine that even today we can not afford. The present total declared U.S. federal debt is about $2.1 trillion. If you add to that the other future obligations that make up the hidden debt (Social Security, civil and military pensions, medicare, etc.) the total is more like $10 trillion. With future budget deficits continuing to add to this figure, it's entirely conceivable that before the year 2000, Americans will be paying up to $600 billion a year in interest payments alone, just to break even. How many politicians have you heard talking about how to solve this problem? Yet politicians continue to borrow and spend and pretend that it won't catch up with us someday.

Our priorities have become drastically distorted. Today many of our children don't have much of a future because we're not giving them a very good start. Roughly half of all school age kids on our planet don't attend school. Of 3.5 billion adults, 880 million of us are illiterate. Some children have no future at all. According to UNICEF estimates, 15 million kids die each year unnecessarily. That's 40,000 a day, 1666 a minute. Four million kids could be saved each year by immunizations costing fifty cents each, or for roughly the amount of money the world spends in one minute on the arms race. Two million die annually because of respiratory infections complicated by malnutrition. Five million who perish from diarrhea could be saved each year by a few cents worth of oral rehydration salts per child. In 1984 the world spent an average of $30,000 per soldier and yet only $455 per child on education.

What we need is a change in vision, a change in social consciousness that allows us to think as a global family and create a more enlightened future because we have learned to believe in ourselves again. Merely by changing our thoughts in the present, we can manifest any number of future worlds. Why not plan a future for our children that we would love to live in now?

Americans and Soviets today are enemies because they have been brainwashed. Why not turn this powerful indoctrination process in a positive direction? Why continue teaching our children fear and mistrust when we can teach them to love one another and build the foundation for lasting peace? Have we not taught them enough about war, killing, and death? Instead of training soldiers, we can raise peacemakers.

Think of the potential that children represent. A newborn child can become anything. Like a computer, his or her mind can be programmed in a million different ways. He or she can be taught to think and speak any language, to believe any ideology, to defend and die for any country. Born in America, he or she may become a national hero, a champion of democracy, even the President. That same child raised in the Soviet Union could just as easily become a prominent communist party member, even the General Secretary. Is there a difference in the man (or woman) or in the child from which he or she came just because of what he or she believes? Only the programming is different. Same biology, same person, different ideology.

The way in which we program our children needs to be updated. Our educational system today largely tries to fill the mind—the container of knowledge—rather than to expand or increase its capacity. And yet psychologists estimate that we use at best only 10- 20 percent of our brain's potential. That's not very much. Try driving your car with only one cylinder working or 15 percent of the brakes. With the other 85 percent of our brain unused, unconscious, or otherwise "out to lunch," it's no wonder we're having problems dealing with each other. Thinking only limited thoughts, perceiving only a narrow portion of reality, using only a small part of our intelligence, we are like the blind men and the elephant—each knows only a small piece of the whole, each limited perception is the truth for that man, yet none knows the greater reality of the elephant.

With his limited perception of himself and the universe, humankind has divided the world into an unlimited number of incomplete truths. Deep within the feelings that we knew as children, we share a common language, yet we have created many hundreds of tongues that hinder communicating with each other from our hearts. We have looked into our souls and the mystery of creation and seen a supreme intelligence, and we have called it many different names. Bowing down to a thousand different gods, we each have thought our creed was the supreme truth and we have been willing to fight and die for it in holy wars to prove others wrong.

And so it has gone, generation after generation after generation. Are we condemned forever to a divided world endlessly battling one another because we are caught up in the drama created by our nearsightedness? Surely, if we could use the full potential of our intelligence we would easily see the futility of continuing to play the game of nuclear nonsense. What if we could expand the conscious capacity of the mind and use 50 percent or even 100 percent of our potential?

Well, we can. Most people have had the "AHA!" experience, where the mind seems to make a sudden leap forward and arrive at an important realization without having to go through the normal process of reasoning to get there. A few have experienced a moment of genius or great intuitive insight in which all the sleeping neurons wake up and connect and out comes something like $E=mc^2$. That was how Einstein did it. Even today, 32 years after his death, his theories are still being proven although he never saw the proof himself.

In the age our children are entering, experiences of higher states of consciousness will be common place. They already are for many. Not only is the consciousness of the planet as a whole accelerating due to the process of evolution of intelligence in general, but we now have available to us a growing technology of consciousness that allows individuals to drastically speed up their own evolution. There is nothing new about this technology, it has always been available, but in the age of ignorance now coming to an end, it has been lost to all but a few. In the next chapter this technology will be explored in more depth.

It certainly makes more sense to teach our children to use their full potential than how to put a bayonet through somebody, for instance. If everyone knew how to experience the unlimited source of peace and tranquility within, fewer people would have to run to the Valium bottle as a stress reduction technique. Instead of spending a thousand billion dollars a year on war toys, how about a mere $10 billion to determine the best techniques for making us nicer to each other?

When we look ahead to the world of our grandchildren, great grandchildren, and great great grandchildren, it is even harder to fathom, and will be vastly different from today's world. Our knowledge is growing at a phenomenal rate. When I went through medical school in the seventies we were told that the sum total of our medical knowledge doubled every 100 years prior to 1900, every 50 years by

Brother and Sister

We are all brothers and sisters here and long-lost friends—although many have forgotten. We are in the process of remembering.

1950, every 20 by 1970 and by the time I graduated it was doubling every 5 years and was predicted to double every year or two in the near future. Look what has happened in the thirty years or so since the discovery of the transistor, yet still our computers are primitive compared to those we will see only 20 years from now. The recent breakthroughs in the field of superconductivity may create a similar technological revolution in how we use electricity. At a recent conference of American physicists, new breakthroughs were being phoned in even as the conference took place. By the year 2000 and probably sooner, anyone will be able to tap into a central computer and have all of mankind's knowledge at his or her fingertips. Because of this, and the expansion of individual consciousness, the evolution of our intelligence and knowledge will be greatly accelerated.

Advancements in the field of health are likely to drastically change the human lifespan. In the fascinating book *Life Extension,* by Durk Pearson and Sandy Shaw written in 1982, the authors summarized our present knowledge about prolonging our lifespan. They present sufficient evidence that leads to the conclusion that humans will live to be 120 years old by the early 21st century and still be healthy, productive, and happy. Even the cure for cancer and other degenerative disease is likely. Perhaps the understanding of the immune system we gain through our search for an AIDS cure will lead us ironically to a solution for cancer that may have eluded us if it hadn't been for AIDS.

With further understanding of the DNA code—the program controlling all bodily functions—genetic engineers should be able to isolate the genes responsible for the aging process itself. It is not too far fetched to envision a time when our great grandchildren will receive routine immunizations that selectively inhibit these aging genes and extend life indefinitely. And for any pessimists out there who insist that mankind is forever doomed because of inborn aggressive tendencies, genetic surgeons in the future should be able to isolate and excise those genes as well.

That our grandchildren will not be earthbound is no longer science fiction. Hypersonic transports, now on the drawingboard, will take off from regular airports, thereby opening space up to the ordinary citizen. Permanently manned space stations and lunar bases will be a reality in the first part of the 21st century. Commercial development of the asteroid belt, permanent colonies on mars, and manned exploration of the outer planets are likely by the mid 2000's. Vast orbiting space colonies capable of growing their own food will be available for our great grandchildren, the new pioneers of humanity. By the latter half of the next millennium, progress on interstellar drives fueled by yet undiscovered forces will allow our descendants to reach out to the stars and contact other intelligent races.

Yet for this vision to become reality, those of us living today will need to solve the age old problem of war and senseless national rivalry that threatens to set our civilization back hundreds, maybe even thousands of years. Just as Americans and Soviets look back to the great men and women in their lineage with reverence, so will future historians judge us by the wisdom we live by today. Are we not, after all, the forefathers and foremothers of tomorrow?

Perhaps we need to look to our children and ask them how to bring about peace on the planet. In their innocence—yet unspoiled by the confusion of their elders—maybe they are closer to the truth than we, their supposed teachers, are.

Do a little experiment to see if you remember the child that still lives in you. Get comfortable and close your eyes for a moment. Do whatever you need to relax. Breath deeply several times and become aware of the feelings and sensations of your body. Imagine what it is like to be a child again—before there was homework, before responsibilities, before adults intruded on your world and spoiled it by demanding that you grow up. Was not the whole world around you a vast playground for endless adventures? Can you remember lying in the cool grass of a hilltop meadow and having nothing to do except watch the clouds turn into unicorns as they drifted above you? How many times did you feel a cool breeze against your cheek or launch hundreds of dandelion seeds into the wind with a puff of air from your lips? In this innocence and simplicity comes peace, because through the eyes of a child we again are in touch with nature, with the life force, with our hearts.

Once, while in Kiev, some women from our group were standing on a residential street corner, lost and wondering which way to go. Before long, a group of children playing across the street noticed them, stopped their game, and got into a huddle. Then, without words, they went to a nearby garden, picked some flowers, and smiling, walked over to the women and gave them their gifts.

In Kiev with a group of Italians, and in Leningrad with a group of Rumanians, we danced arm in arm and sang this familiar song by Lionel Ritchie with each other:

Dennis L. Haughton, M.D.

"We are the world, we are the children,
We are the ones who make a brighter day,
So let's start giving."

We barely spoke each other's language or knew one another's customs, yet in the jubilation of living that moment in the here-and-now together, it barely mattered. By getting away from the complexities of world politics and playing together we created the natural conditions for peace to arise.

Through the eyes of a child we will see the path to peace. It was Jesus who told us that the Kingdom of God belongs to people who have hearts as trusting as little children. We are all just children inside beneath our battlescars from living in a troubled world. We all need much more love than we usually get. Deep inside, we are all brothers and sisters and long-lost friends. Very soon, we will be coming home again and realizing the wonderland paradise we can build all around us in which to play.

CHAPTER EIGHT

Ahead to the garden

The present age is coming to an end and a new one is beginning. We live now in a time of rapid transition. Humanity has lived for thousands of years in an epoch that future historians may very well label the Age of Ignorance in comparison to the era that we are entering. Prevailing social consciousness has been dominated by the belief that everything life has to offer is present on the obvious levels of existence, and that to aspire to anything that might lie deeper than external appearance would be useless. From this superficial understanding of life, insight into more fundamental levels of truth has been lost and the natural harmony with nature and the life force has been forgotten. Tension, confusion, unhappiness, fear, war, and suffering have reigned for centuries because of this limited view of life.

Humanity, in essence, has been suffering from an illness of the soul. We have lost our awareness of and connection to the very essence of life—whether that be called God, the Lifeforce, the Tao, the field of pure consciousness, or any of the multitude of other words we have named this something beyond words. We have become slaves to many lesser gods: money, power, sex, fame, drugs, and material possessions.

Because this most basic connection is missing, we also suffer from an illness of the heart. We have forgotten how to love. We all yearn for love but think it is found somewhere outside of ourselves. We never will find it out there until we find it first inside. It is the unconditional love of self that we lack most, for from this fullness arises the power to truly love others without limitations. At the basis of this lies our difficulty in accepting ourselves unconditionally—beyond the judgement of good and evil—for all that we are. As our hearts open and we find acceptance, we will learn to forgive ourselves and stop judging and condemning who we are. From this forgiveness of self will flow self-love which will allow us to forgive each other for our collective misdeeds. A lot of love is needed to heal the individual and collective wounds of our planet. Instead of condemnation and harsh criticism, we will learn to reach out across the oceans with love and good will as we reconnect.

We have also been affected by a pervasive illness of the mind and intellect that has colored our perception of life with pessimism. It is as though all along we have been wearing a pair of glasses that tints what we see with doom and gloom, and yet we don't know that we wear them. If you find it difficult to imagine a world at peace with itself, perhaps you have been submerged too deeply in the world of TV and the media's fascination with tragedy and suffering. Maybe you have been wrapped up in the mundane details of daily living and have not spent enough time listening to the wind or gazing up at the stars. Our limited use of our mind has created a world artificially divided where fear, judgement, and mistrust abound. Perhaps the boogyman we have created in our unconscious from our collective terrors scares us from looking deep within and finding the wisdom to heal that also dwells there.

Because of the underlying lack of fulfillment on the level of heart, mind, and soul, our physical bodies have become battlegrounds of disease that are a reflection and consequence of the lack of wholeness. Illness comes about not through the action of germs alone, for they require a weakened state of immunity to take root and grow. The absence of disease doesn't create health, wholeness does.

As we move toward individual and global wholeness we are beginning to understand life from a much more profound level. Until recently, the prevailing system of gaining knowledge about something has been through objective observation and intellectual analysis—a predominantly left-brain function. This is the fundamental basis for the scientific method. Knowledge gained in this way, however, is incomplete because it leaves out the vast intuitive capacity of the right brain and ignores the subjective state of the person gaining the knowledge. Einstein's breakthroughs came largely through his intuition, his logical mind caught up later. How can a person claim to know the reality of a flower if he has no idea of who *he* is. Thus, complete knowledge is possible only after the integration of right and left hemispheres takes place and the person knows himself as well as the flower outside of himself. The new age is coming about largely through the process of looking inward.

What we are discovering right under our noses is that our greatest natural resource, the mind, has largely been untapped. Once relegated to the mysterious world of mystics and the recluse living in solitude, the study of the mind by turning our attention inward has evolved into a science of consciousness that anyone can learn.

The technology of consciousness, or science dealing with the subjective mind, has been around for as long as we have, only it has been buried in the background in the Age of Ignorance. Throughout recorded history, there have been individuals whose insight, wisdom, and intuition have far surpassed ordinary human consciousness. In 1901, Richard Bucke, a Canadian physician, made a scientific study of the evolution of the human mind in his book *Cosmic Consciousness*. He saw the individuals who reflected a higher level of consciousness such as Buddha, Jesus, Paul, Blake, and a growing number of others, were just forerunners of the beings who would eventually people the planet.

Actually, there really is nothing supernatural or mystical about higher states of human consciousness. Life, after all, is but the evolution of intelligence toward ever increasing complexity. Isn't it reasonable to believe that people will probably not evolve wings or three eyes but a more powerful mind? Today, scientific research into higher levels of consciousness has been extensive. Not only can the internal subjective mental state of a given level of consciousness be well defined, but the physiological and biochemical correlates are quite specific for each one.

Bucke, in his book, defined three basic states of consciousness. Simple consciousness is the awareness of the external world and inner bodily sensations perceived through the senses as typified in the animal kingdom. We have taken this one step further in self consciousness, in which we have become aware of ourselves as distinct thinking entities apart from the rest of the universe. From this is born contemplative thought, language, culture, and the pursuit of ideas and knowledge for their own sake. Finally Bucke goes on to describe the goal of human evolution, cosmic consciousness, as an awareness of—and ever-present conscious connection to—the universe as a whole, in addition to self awareness. This places the individual on a new plane of existence as far above self consciousness as we are presently above animals. The characteristics of this enlightened state he describes fully through the teachings and writings of individuals throughout history that expressed it. Eventually, he concluded, our descendants would all evolve to that plane.

In a more modern framework, we all have experienced the three ordinary states of consciousness: waking, dreaming, and sleeping. In meditation we experience a fourth major state: transcendental consciousness. This can be defined as the internal awareness of pure consciousness itself, a state beyond thought, sensation, and external awareness in which the invisible fabric of the mind itself is cognized. Intellectual understanding alone can never suffice to give complete knowledge of the higher states of consciousness: they have to be experienced. Through meditation, I have encountered pure consciousness countless numbers of times and all I can say is that the experience is one of inner tranquility, peace, and wholeness that is beyond words to describe.

Our knowledge about anything outside of ourselves depends on our internal state of consciousness, or in other words, knowledge is structured in consciousness. A room containing a bed, mirrors, windows, and houseplants completely ceases to exist in sleep and is filled with thrashing six-headed monsters in the dream state of consciousness. Likewise, when the experience of pure consciousness begins to grow in the higher states, our perception of reality expands. Problems that once appeared insoluble simply vanish as layers of ignorance are peeled away. In the light of day, the woods—which in darkness contained all sorts of goblins and trolls—becomes a beautiful meadow with rabbits and wildflowers. The earth which once was limited and flat suddenly became round and unbounded when Columbus sailed back from the new world without falling off the edge. By cultivating the source of peace within, our doom and gloom glasses are removed and we can see solutions to age old problems that have been there all along but have been hidden by our ignorance.

Changing the outside is accomplished by first changing the inside. By changing consciousness itself, we are acting on the most fundamental level of existence and therefore the effect will be the most profound. The world is currently living out a racial megascript that we have first created in our collective consciousness. Our belief in our craziness inside creates the manifestation of it on the outside. Just by changing our consciousness and our thoughts we can reprogram a new magascript and create any type of world we choose.

Global awakening needn't take centuries, it conceivably could happen overnight. What do you think would happen if we had a world wide election to determine whether or not we would have war on the planet any more? What if everyone including the children (after all, it is their future) got one vote. Governments had no say in the matter except that each official could cast his or her vote. Do you doubt what we would decide? The point is, if 90 percent of us desire peace, what is to prevent us from making it a permanent reality?

There are some striking parallels between the science of consciousness—the supreme expression of subjective knowledge gained through the intuitive right brain—and quantum field theory—the supreme understanding of objective science gained through the logical deductions of the left brain and scientific experimentation. Physicists, in perusing the belief that all matter and energy arise from one fundamental underlying field, have described a vacuum state—devoid of any measurable energy or matter—from which an unlimited number of particles come into being and vanish without end. Through meditation by simply turning the attention inward, masters of mental exploration have, through their subjective search, also found that all thought and external reality ultimately arise from the field of pure consciousness.

No longer is the science of consciousness considered wishful speculation. To my knowledge, to date the most complete research on consciousness has been carried out on practitioners of the Transcendental Meditation technique introduced to the west by Maharishi Mahesh Yogi almost 30 years ago. I myself have meditated regularly since 1971. I have taken many of the advanced courses including instruction in techniques that, among other things, teach an individual to levitate above the ground by a simple thought.

The most fascinating studies that I have read involve the changes in brain waves seen in advanced meditators. Normally there is very little coordination between brain waves recorded on an EEG between the front and back and the right and left hemispheres of the brain. However, during liftoff in the levitation technique, a unique thing happens: suddenly the brain waves between front and back and right and left become synchronized and pulsate in unison. In this state of syrchromy the brain functions as a whole, something not witnessed before in humans. I can only say that the energy I have felt traveling up my spine during this experience is not only powerful, but very exhilarating.

Once having gained familiarity with pure consciousness through regular meditation, advanced techniques called Sidhis allow a person to begin programming the cosmic computer inside to create a whole range of effects on the outside. The point in discussing the science of consciousness here is that today we have mental technologies available to us that can allow anyone to use much more of his or her potential. We no longer have to wait until random circumstances spontaneously produce isolated individuals exhibiting higher states of consciousness. We now have the opportunity of systematically using this knowledge to accelerate the growth of consciousness exponentially, thus speeding up the evolutionary leap we will eventually take. In addition to mediation, numerous other techniques—some purely mental and others using advanced technology—are showing promising results in expanding consciousness and tapping greater potentials of our brain.

Research into consciousness has revealed other information with profound sociological and global implications. The effects generated by some of these advanced techniques are seen not only in the individuals practicing them, but they seem to induce more harmonious and orderly behavior in persons far removed who have nothing to do with consciousness enhancing techniques themselves. Scientists associated with the research claim there is a field of infinite correlation through which everything in existence is connected. By stimulating this level at one point, ripples extend outward to affect everything else.

The effect is not just hypothetical. Various scientific studies have demonstrated that when a very small percentage of individuals enliven the level of infinite correlation through regular practice of these techniques, such things as accidents, crime rates, and other measures of social disorder decline in the population as a whole. Some of the more striking experiments in the late 70's involved sending large teams of individuals trained in the more advanced techniques into various trouble spots in the world that were undergoing violent conflict including Iran, Nicaragua, and El Salvadore. A day or two after the teams arrived, levels of bombings, killings, and other measurable acts of war dropped dramatically and remained low during the weeks they were there. The teams did nothing more except sit in their hotels and meditate, using their advanced techniques. Several days after they left, the hostilities rose to their prior levels. As it turns out we are far more interconnected than we once thought.

A phenomenon that occurs in nature has profound implications for world consciousness as a whole. In a system such as a laser, when a very small percentage of atoms has been induced to vibrate in unison, there is a sudden phase transition in which all the remaining atoms are raised to the same level of excitation, and the whole system achieves perfect order.

There is a true story told by Ken Keyes, Jr. in his book *The Hundreth Monkey* that demonstrates how this principle applies to socio-biological systems. In 1952 on the island of Koshima near Japan an 18-month-old female monkey named Imo learned to wash sand off her sweet potatoes before eating them. Over the next six years many other monkeys learned to wash their potatoes by following her example. Then in 1958 when a certain critical number of monkeys had learned this trait, something very surprising happened. Suddenly, nearly everyone in the tribe started washing their sweet potatoes. Even more surprising was that colonies of monkeys on other islands and the mainland, who had no direct contact with Imo's tribe, also suddenly began washing *their* sweet potatoes. Thus, when a certain critical number achieves an awareness, this new awareness may be communicated from mind to mind.

The same apparently holds true for human systems as well. Originally called the "one percent effect" by TM researchers, because it took one percent of the population meditating to demonstrate the effects, it since has been hypothecized to occur when only the square root of one percent practice the more advanced techniques. Not only can a small number of individuals functioning from a more orderly level influence the rest of the world, but as world consciousness as a whole rises, a positive feedback loop creates more profound internal experiences in the individuals meditating. The main point is that the transition between one phase and the next, as in the laser or the monkey experience, happens suddenly when a certain critical threshold has been reached. The world is ripe for an evolutionary leap that could catapult us to a higher order of existence, not in thousands or millions of years, but conceivably in a single generation or even in months, weeks, or days.

Although the science of consciousness has been around for thousands of years, widespread Western-style scientific research into these states has only been around a few decades. As we gain more experience tapping the powers of the mind, we will be able to refine our knowledge and accelerate our scientific understanding. Then, through practical application, we will be able to hasten the emerging Age of Enlightenment. Although impossible to predict the exact time frame in which this will occur, I believe the exponential growth portion on the phase transition curve is imminent. Mikhail Gorbachev's recent policy of glasnost is a good example of this rising world consciousness. How quickly we make the transition depends a lot on our underlying attitudes and how much we consciously move toward global wholeness.

It is now time to stop manifesting the dark side of the life-force and move on to better things. Becoming whole means the successful integration of all our cultures and ideas. Each of us represents only a part of the whole. That doesn't mean that we must give up our individual uniqueness and conform to one ideal. Diversity is important in creation. Hundreds of different trees and plants live harmoniously together to make a forest. Without the north and south poles a magnet wouldn't exist. Without darkness there would be no light. Democracy and communism just represent two sides of humanity's experience in creating social order.

It is time for nations to put aside their weapons, propaganda machines, and denunciations of each other. It is time to reap the harvest of the collective wisdom we have gained through living together—while we were divided from each other—and build a world that we all desire. We now have instant global communication, what we lack is global understanding. It will come. In the very near future, certainly before the year 2000, the language barrier will finally fall as personal electronic translating devices, capable through speech recognition and synthesis circuitry, will make communicating with anyone in a foreign language as easy as talking. Already, handheld keyboard models are available. Beyond that, telepathy is a distinct possibility when, instead of using only 10 or 20 percent of our brain, we learn to use 30, 50 or even 100 percent.

Already, psychic experiences have become commonplace in ordinary people. Shirley MacLaine and Richard Bach are just two of my favorite authors who have written of their own experiences. While in the Soviet Union I was amazed at the level of communication we were able to establish with one another using very few words. Touch and eye contact were often enough to establish heart-to-heart understanding. The most profound awareness that came to me during my visit to the land of the "enemy" was that, beneath the misunderstanding, fear, and mistrust, there was a bond of love between us that connects us more strongly than our ideologies repel us. Instead of nuclear missiles and the threat of genocide, peace in the future will be assured by a network of global consciousness connecting us with bridges of love.

There is a growing need for a world government of the people, by the people, and for the people: all of us. Again, this doesn't mean the destruction of any individual government currently in power. For the benefit of all, we need a higher body of authority composed of the best features from all our governments to oversee planetary concerns and settle international disputes when they arise. Some sort of an international peacekeeping force, controlled by the people and not the other way around, would replace most of the monstrous armies and war machines we currently maintain to wage war on one another. I don't think all our nuclear weapons should be destroyed either, but that some of them be held in trusted hands for peaceful uses. It would be nice to have some high explosives sitting around when earth monitors, for instance, detected a major asteroid heading our way which is quite possible in the years to come, according to some scientists. After all, the dinosaurs were made extinct in the aftermath of a collision with some wandering extraterrestrial body millions of years ago.

As we grow up and become whole we will likely move on to a state of unprecedented equality. All people, no matter how rich, poor, famous, or unknown—and regardless of race, creed, religion, sex, color, political party, height, eye color, and so on—will be treated equally. In the awareness of our connection to the greater life flow, there are none that are more than and none less than—we are all just people. Not only the United Staters constitution, but the Soviet constitution of 1977 as well, aspires to these ideals. Mikhail Gorbachev certainly seems to be advancing his nation in that direction. In a recent TV program I heard Vladimir Pozner reflect on his culture's goal of creating a society that one day will need no government at all and where there will be no oppression. When you think of it, it really is no surprise that beneath the superficial aspect of national ideology, there is a more fundamental body of ideals shared by us all.

Experiences with higher states of consciousness will abound in the coming age. After all, it is the rise in consciousness that is catalyzing the changes already occurring. We have beings in our midst now whose wisdom completely shakes the limited vision of who we think we are. Channeling through a woman named J.Z. Knight, Ramtha, a being claiming to be 35,000 years old, has been speaking to audiences since 1978 when he materialized in front of a startled Mrs. Knight and her husband, while she was building pyramids in her home one weekend. I have listened to his wisdom, both in person and on over 30 hours of videotape, and his teachings are profound. Despite skeptical critics who seem more interested in the more superficial aspects of Ramtha and J.Z., I don't think there is any way, either through fakery or any type of trance, that a common housewife from Washington could be making up the wisdom and knowledge that flows as Ramtha speaks. His reason for being here with us at this time is to help illuminate our path into the coming age. I believe that once we have finally gotten our own act together, we will be ready for open contact with extraterrestrials, who are also part of the same life flow. The dim view that the earth is the center of the universe and mankind is the only intelligent lifeform in creation died long ago. Knowing what we do now about the development of life from simple organic molecules to intelligent lifeforms, it is a near certainty that intelligent beings like us exist all over our galaxy, to mention nothing of the infinite galaxies beyond.

In a limited way, we may already have been visited and continue to be visited by E.T.'s. The reason they remain for the most part hidden, I believe, is twofold. First, I think they respect our right to evolve on our own and arrive at racial wholeness or destruction through our own wisdom, without outside intervention. Secondly, it's probably too dangerous here, considering the barbaric way we still settle our differences by killing each other. What intelligent extraterrestrial would want to go to New York City only to be blown away by a mugger on the subway? When open contact is established, sooner or later, then our collective consciousness will expand a hundredfold and our vision of possibilities will become unbounded. For then humanity will become part of a network of interconnected centers of intelligence that stretch onward to infinity.

We are now at a crucial juncture in our evolution as a race. I believe that the popular support for creating world peace is already with us, yet is untapped. The world looks expectantly to the super powers to see how they will settle their differences, for if they can overcome their ideological obstacles and establish good will toward one another then the rest of the world can quickly follow. The Soviet Union and the United States have a golden opportunity before them to lead the world toward global peace. I make the following proposal to Mr. Reagan and Mr. Gorbachev as they prepare for their next summit meeting:

1. That, together, Americans and Soviets consider peace, not from the limited viewpoint of just reducing nuclear missiles and conventional armaments, but from the standpoint of creating planetary peace and wholeness as viewed from our most enlightened perspective.

2. That, together, U.S. and Soviet heads of state make the bold decision to end once and for all the cold war between us and lead their peoples to friendship and the rest of the world toward demilitarization and global peace.

3. That misinformation, denunciations, and any other form of official propaganda intended to create fear and mistrust between countries cease forthwith.

4. That funding for a joint peace initiative be taken from current defense department budgets.

5. That citizens from both countries be given widespread opportunities for meeting face to face to communicate about our mutual concerns and to establish bonds of friendship and goodwill that will provide the solid basis for ensuring lasting peace.

That's it. That is my prescription for the illness facing our planet. But what about you? Is there anything you can do to speed up the healing process? You bet there is. The possibilities are endless and limited only by your imagination. I ended up writing a book. In the global awakening taking place we can all become peace makers, but we must begin by looking inward and first making peace there. If each of us takes responsibility for the piece of the planet that we live in, the rest will follow. Just envision what you desire and follow your heart.

* * * * *

There is a symbolic story from the Bible that applies to all of us:

In the beginning, God made man and placed him in a beautiful garden called Eden. And the Lord God planted all sorts of beautiful trees there which produced the choicest of

Hauasu Falls - Grand Canyon

Our world is a garden paradise and we, its caretakers. In the days to come even the deserts will be made green with the abundant waters of the seas.

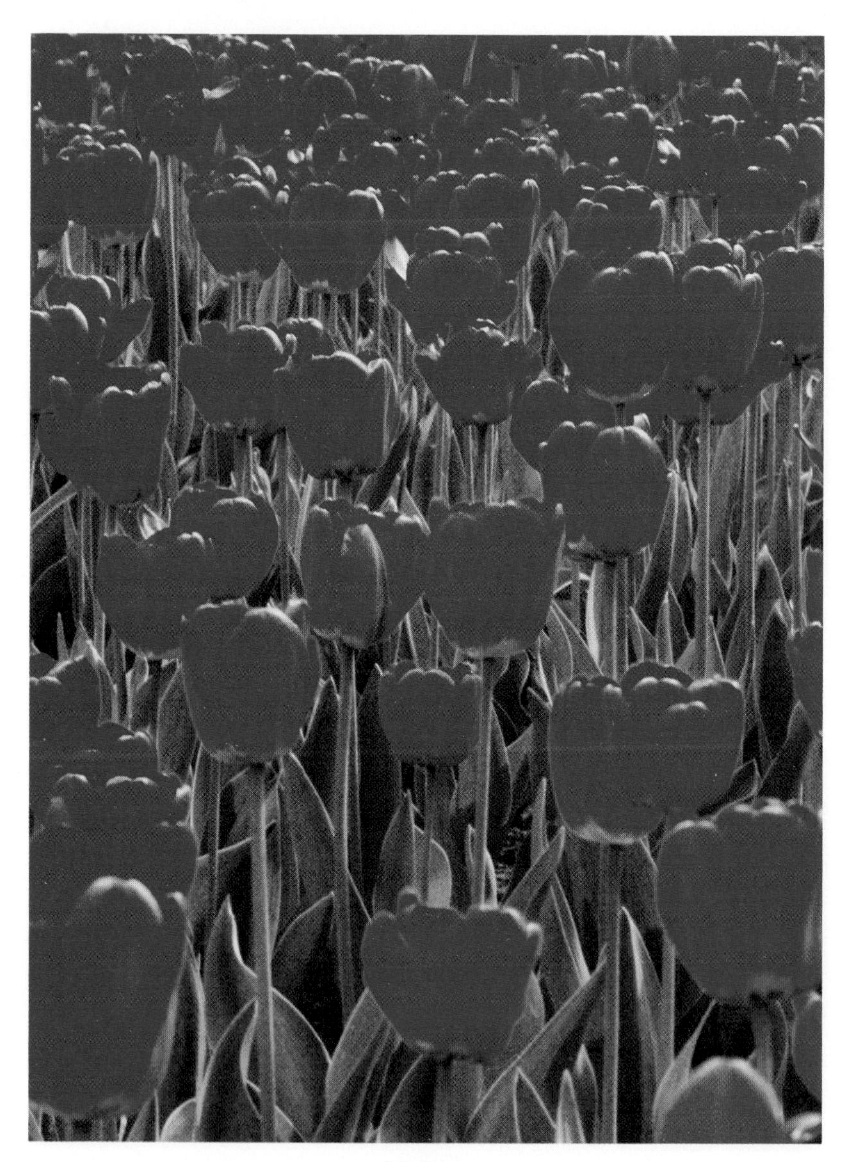

Tulips

Are these communist tulips or were they raised in a democracy? Of all the lifeforms on the planet, only humanity has divided things into good and evil.

fruits. In the center of the garden he placed the Tree of Knowledge. And the Lord God told man that he could eat from every fruit in the garden except the Tree of Knowledge—for its fruit would open his eyes and make him aware of good and bad, right and wrong. And He told man that if he ate of its fruit, he would be doomed to die.

From the time we ate of that fruit countless eons ago, we have been lost in the world of duality, of right or wrong, good or evil. There is a place beyond duality, beyond life or death that we are approaching in our awareness. Duality and judgement are but illusions that have hidden the unbounded limitless reaches of forever from our vision. It is this that has kept us in ignorance of the completeness of life surrounding us.

Is a yellow bird right or wrong, the spring good or bad?

The lilies of the field . . .

Falling leaves . . .

Are they bad? Is one more right than another?

Nature just is. It flows in a stream called life that unfolds beyond the duality of judgement. Man too, with all his beliefs that he alone has judged, is part of that same life flow. You see, we have never left the Garden, for it has been around us forever. It has only been our ignorance that has separated us from it.

Look at it shining down there. Isn't this world with its cool blue oceans, its puffy clouds and gentle rains, its fresh spring daffodils and rolling green hills a splendid place? Amongst billions of worlds in our galaxy, only a few could be as nice as this. We couldn't have been given a better home. Even the brown we will turn to green in the days to come with the abundant waters of the seas. We have been given the complete freedom to make it the paradise of our dreams or destroy it forever in an instant. The choice and the power for both are ours.

To all of you down there, you are my brothers and sisters, yet many of you don't know that yet. I extend my peace and good will to you all.

I, for one, vote that we move our world onward into the garden we thought we had lost eons ago. What about you?

EPILOG

Cultivating the new awareness

"The splitting of the atom has changed everything, save our way of thinking, and thus we drift towards unparalleled catastrophe. We shall require a substantially new manner of thinking...to survive."
Albert Einstein

Forty two years have passed since the Bomb was dropped. The fact that it happened to be an American bomb and a Japanese city matter very little now, for sooner or later someone would have played the role and dropped it on someone else. And for 42 years since then we have been struggling under the immense psychic burden which accompanied such a possession—trying in vain to somehow fit this ultimate weapon into our primitive eye-for-an-eye jungle way of thinking. In the process, we have amassed a horrendous pile of costly and unusable weapons which even the least educated among us is beginning to recognize as a symptom of runaway racial folly.

Yet, despite our struggle on the brink of catastrophe, we have survived. Along with the means for our extermination, Dr. Einstein also gave us a profound mental challenge that is literally forcing us to develop a new manner of thinking in the span of a single generation. In short, we have contemplated the awesome power of the Bomb and are gaining the wisdom that, in the nuclear age global war has been rendered obsolete, for there would be no winners. The Military Age, in which we have settled our disputes on the battlefield, is coming to an end. Even conventional war between the superpowers—which can always escalate into global thermonuclear holocaust whether or not we are currently stockpiling the Bomb—is becoming a no-win alternative. World War III will never occur.

However, when we look around us today, examples of the old style of thinking still seem to predominate. The new awareness that we are moving into so far is still in its infancy in our social consciousness. Many have not perceived it yet, and yet many of you are beginning to. It is my sincere desire that, wherever you are, you have felt the stirrings of a new hope, a new optimism for the future, a new knowingness growing in your heart as you have read this book, because that is the beginning of the new awareness awakening within you. As you begin to feel this inner change—from the old style of thinking that created powerlessness ("Lasting peace is impossible.") to the new that empowers ("Of *course* we can create peace if we desire it")—then you have begun to see how easily peace can come about. It is only a matter of a mental shift—a change in consciousness within. This is the first and most important step, and it can occur in an instant.

Once this inner change in awareness has taken place, the seed has been planted. The next step is to nurture it, to cultivate it, and together create in our individual awareness and in our social consciousness a new manner of thinking. If we think "war" we will create war, think "enemy" and we shall make enemies. Yet think "friend" and we beget friends, think "peace" and we shall create peace. You see, global peace depends on all of us who live here, not just on our leaders and diplomats. We, the people—the children of Eden—are the peacemakers. Each of us can play a role in fostering this new manner of thinking so that we can harvest the fruit of this awareness—global peace—as quickly as possible. The following are some of the things you can do to help.

First, make peace within. This is the most important and most powerful thing you can do on this planet, not only to create global peace, but to achieve the fulfillment of your own desires as well. The biggest war we all face is the turmoil inside which hides from our view the unlimited storehouse of tranquility that lies beneath it. I wish I could take you there, but only you can do it for yourself. There are many pathways to inner peace to choose from, and I have included a few sources at the end of this book for you to explore. I prefer meditation (TM) because I have practiced it for 16 years, but also because it is easy, gets immediate results, is taught the same way worldwide, and is the quickest and most direct route to begin tapping the field of pure consciousness within that I know of. Another pathway may fit you better, however.

Learn to love unconditionally, especially yourself. The force that is bringing about planetary healing and mending the wounds that separate us is love. Fear, mistrust, accusation, hatred, rivalry, conflict, and disharmony have dominated international relations with each other until now. By learning acceptance, forgiveness and unconditional love for ourselves, we will automatically find tolerance, trust, goodwill, cooperation, and brotherly/sisterly love growing for each other.

Become as children again. Be good to yourself. Slow down. Relax more. Play. Lie on a grassy hilltop watching the clouds drift by and daydream about the new world we can enjoy together by joining hands instead of throwing bombs. Learn to lighten up on yourself and one another. After all, does it really matter very much that we don't all look and dress the same and share the same beliefs? There is room on our world for everybody no matter how different. And do you spend all your time working for some future goal when you could play more today and be happy in the here and now? When you find your grownup battlescars overshadowing your new awareness with hopelessness and pessimism, ask your children and listen to their wisdom. Their unbounded imaginations may give you new ideas and restore your faith.

Think globally. Begin thinking of yourself as a world citizen and member of a global family, for are we not all descendants of the children of Eden? Pretend that everyone you see is a distant cousin, and aunt, or an uncle, because they are. It will surprise you how this simple mental shift changes your perception of others. As you begin feeling the connection that joins us together, you will find yourself becoming more tolerant, more understanding, and more forgiving of others. When we begin to see this Spaceship Earth as our home we will become more loving in how we care for it and our fellow passengers.

Reach out and connect with others. Remember Hands across America? Why not Hands Around The World? American and Soviet citizens have been linked by many different space bridges to encourage peaceful exchanges and understanding with encouraging results. There is the sister cities program which has paired Soviet and American cities to foster citizen to citizen contact. Why not promote more widespread exchanges like this? We have the technology to do so easily, all it takes is the desire to bring it about. My sister for example, is developing a program to connect Soviet and American highschool students together via computer link-up.

Start a peace chain letter. In October 1987 just such a letter is being launched in the Soviet Union and the United States simultaneously. Each person in the chain will get two friends to write a declaration of peace to Mr. Reagan and Mr. Gorbachev asserting our common desire as people to put an end to war and live peacefully with each other on the planet. Each generation of the letter will double the number of people involved. After 30 steps, more than 2 billion of us will be united and sending letters. A billion letters would form a cubic pile 100 feet on each side, hardly something that can be ignored. And all it took was an idea and a couple postage stamps to start. You can begin your own.

Learn to recognize propaganda and help stamp it out. Anything exalting one side by condemning the other ("We're good—they're bad.") should be immediately suspect. To become better informed go beyond the usual media sources of information no matter where you live. Recognize that each culture has its own point of view and goes to varying degrees to promote it. Demand from your political leaders, whether they are in the U.S., Soviet Union, Iran, Iraq, Nicaragua, or anywhere else, that they adopt a rational, humanistic, pro-life, peaceful, respectful, loving, and truthful attitude toward one another so that we can befriend each other and get on with the business of creating a peaceful world.

Become a citizen diplomat. Travel to the land of your "enemy," wherever that may be, and make friends. Find out for yourself that beneath all the political rhetoric, people are people, and that the overwhelming majority wish to coexist and live peacefully with one another on this planet. Americans, put away your fears and go meet the Soviets. They, like us, are wonderful people. Soviets, convince your leaders to lighten up some and relax their travel restrictions. Come to my country and see for yourself that Americans too are good loving people—we really don't all carry guns here. If you can't afford to travel, organize a raffle in which all your friends chip in and the winner becomes the citizen diplomat for the whole group.

Use your imagination and be creative. You will find there are endless ways to cultivate this growing new awareness and thus hasten the full bloom of global peace. Soviet and American school kids who have become penpals send each other wonderful drawings portraying themes of peaceful U.S. - Soviet relations that should be inspirations for our diplomats. A friend of mine in Washington is publishing a 1988 calendar entitled "Children of Peace" which features the faces of some of the Soviet kids he and others have captured on film. Why not create a "THINK PEACE" bumper sticker and become rich in the process. A friend in Phoenix recently discovered a peace message in the nearby mountains that had originated in Australia and had traveled the last leg of the trip by balloon. Whatever you do, have fun doing it and above all know and have faith that whatever small contribution you make does indeed make a difference.

Beyond that, the next chapter is up to you, and me, and all of us everywhere—because what we make of this planet depends on what we desire. As for myself, I am planning another visit to the Soviet Union next spring and perhaps an earlier trip as well to work on the production of a Soviet edition for this book. I believe it is vitally important to our future that Soviets and Americans look beyond the limited goals of arms reduction and begin talking openly about a much more enlightened concept of peace, based not on military deterrence, but on mutual trust, open communication, cooperation, and goodwill. The majority of Soviets I have met are already thinking along these lines. In fact, there is even a new toast circulating in Moscow which translated, means "To no borders." To stimulate this dialog, I explored the idea of a Russian translation for this book on my recent trip several months ago, and found two enthusiastic and influential people in Moscow who agreed to be my liaison with Soviet publishers. In a stroke of good fortune, they both made their first visit to the U.S. a month ago and I was able to meet them in San Francisco.

It has been three months since my last Soviet trip and it was even more powerful than the other. Signs of glasnost and the people's enthusiasm for it were evident everywhere. Now I am even more optimistic for our future than I was a year ago. The wounds from Chernobyl still haunt people's awareness, and because of intense citizen protest, construction of unit five and six at the Chernobyl power complex has been abandoned. While in Tbilisi, I learned that beaches on the nearby Black Sea were still closed due to radiation from Chernobyl, over 500 miles distant.

I will never forget the feeling of celebrating the Forth of July in Moscow at a rock concert marking the end of a 400 mile joint Soviet-American Peace March and featuring James Taylor, Bonnie Raite, Carlos Santana, the Doobie Brothers, and several Soviet groups. It was an eerie feeling sitting there and knowing that U.S. missiles were targeted on us the whole time. The following day, reflecting a spirit of goodwill common in my travels, a maid in my hotel, seeing my peace march T-shirt, handed me a donation roughly equal to her day's wages to help bring Soviets to the U.S. next year for a second peace march.

Yet by far the most profound experiences I had were in encounter type workshops with Soviet psychologists, educators, and professionals that we conducted in every city we visited. Approaching each other not as adversaries but as people working together toward common goals, we explored many techniques for conflict resolution and bridging misunderstandings between Soviets and Americans. Facing each other openly and communicating on a gut-level, we found that ideological differences were unimportant. The late pioneer of humanistic psychology, Dr. Carl Rogers, before his untimely death set a precedent for this type of work last year by conducting workshops with many of the same people I met on this trip.

During one particularly intimate session in Tbilisi held in someone's home, we began our encounter with a group meditation, led by the Soviets. The people in this group had been meeting regularly for several years and usually began each weekly session with a meditation in which they sent their love to all the Americans they had met before. Several days later we joined 13 of our new Georgian friends for a day-long hike in the mountains surrounded by the beauty of nature that we all recognized as home. While walking through the woods one of the Soviet men said that we, and all people the world over who are jointly working towards peace, unite and be known as the Spiritual Army. I believe there is much our political leaders could learn in this way by spending time together, not at the formality of the conference table, but in the simplicity and quiet of nature.

It is, in fact, nature that connects us, despite the illusion of our separateness. The life force that flows through us and everything else in creation is the same, regardless of our beliefs or ideologies. The new awareness and manner of thinking now coming into being as a result of the evolutionary leap now upon us, is simply a higher expression of this force. I don't believe there is any turning back, because it is not only our actions that propel us forward but the awesome power of the life force itself. It is our unavoidable destiny as human beings to become something greater.

As if on cue, and unknown to me as I wrote this book, August 16 and 17, 1987 marked a fundamental shift in planetary consciousness that had been predicted for the same precise time by numerous ancient phophecies including Aztec, Mayan, and Hopi. This Harmonic Convergence, as it has become known, marks the end of a 5000 year cycle of fear and destruction and the beginning of a spiritually enlightened one. On these dates, thousands of people worldwide gathered at various sacred sites in order to create a collective vision of peace and harmony and a planetary field of trust to catalyze the shift in global awareness.

Jose Arquelles, one of the main organizers for the event, says that Harmonic Convergence is a trigger for a shift in human mental orientation "from a collective determination to view things from a perspective of conflict to a collective determination to view things from a perspective of cooperation." This paradigm shift will be characterized by the end of the arms race, demilitarization, the end to pollution, and the return to environmental harmony.

Yet for some of you, the idea of lasting global peace still remains an impossible fantasy, and the concepts of planetary consciousness and evolutionary leaps are nothing more than metaphysical nonsense. For you, the old way of doing business with each other is here to stay. So be it. I respect your pessimistic doubts and I think they are important to consider. Yet I wonder if this skepticism is not just the biased reflections based on decades of doom and gloom type thinking that has led not only to the present arms race but, when projected into the future, has created the need for a complex Star-Wars defense system to protect ourselves from each other. I believe that this pessimistic age is drawing to a close, and that within a few years, even the skeptics among us will recognize the shift in our collective attitude.

And to those of you who are beginning to sense this wakening new awareness, welcome to a new world that is just beginning to unfold. Though we may not be linked by visible means, we, nonetheless, are always connected in consciousness. Wherever you are on this planet, we shall work in harmony as one to create a new destiny for humankind as we together travel the path to peace. Our first priority is to do away with our immense piles of weapons which could prove to be downright embarrassing if our civilized kindred from the stars visit us before we have beaten all our swords into plowshares. As we continue to connect through bridges of consciousness spanning the oceans and as our global family matures, there will come a day when we, too, will reach out toward the stars and connect with a network of life and intelligence that stretches on to forever.

RESOURCES

The following is a list of valuable organizations you may wish to contact or become involved with.

ACCESS.
1755 Massachusetts Avenue, N.W.
Suite 501
Washington, D.C. 20036

Works with over three hundred experts, national organizations, and research centers to provide authoritative information on issues of peace and international security.

The Association of Humanistic Psychology.
325 Ninth Street
San Francisco, CA 94103
(415) 626-2375

Founded in 1961, AHP concerns itself with self-actualization, humanistic psychology, personal growth, consciousness expansion, creativity, holistic healing, and working toward a society that fosters the growth of fully developed human beings. They are very much concerned with world peace, and their Summer, 1984, issue of the *Journal of Humanistic Psychology,* devoted entirely to peace issues, is quite enlightening.

Beyond War Foundation.
222 High Street
Palo Alto, CA 94301-1097
(415) 328-7756

Its monthly newsletter *On Beyond War* fosters a new way of thinking - based on the fact that we all live on the same planet as one human family - that will allow us to make war a thing of the past.

Citizen Diplomacy, Inc.
P.O. Box 9077
La Jolla, CA 92038

A grassroots clearinghouse on the more than one hundred twenty organizations in the United States involved in starting or maintaining sister city programs between the U.S. and the Soviet Union. It publishes *Citizen Diplomacy.*

Coalition for a New Foreign Policy.
712 G Street, S.E.
Washington, D.C. 20003

An interfaith group working on legislation and providing education in issues.

Grassroots Peace Directory.
P.O. Box 203
Pomfret, CT 06258

Maintains a current list of over eight thousand peace organizations in the United States.

Institute for Soviet American Relations
1608 New Hampshire Avenue, N.W.
Washington, D.C. 20009

Publishes magazine *Surviving Together* three times a year containing articles about what's happening in both countries with U.S. - Soviet relations along with hundreds of short descriptions of cooperative ventures between the two countries that have been recently announced.

Institute for Space and Security Studies.
7833 C Street
Chesapeake Beach, MD 20732
(301) 855-4600

Its newsletter *Space and Security News* is an excellent source of information on nuclear arms issues and SDI (Star Wars). Founded by Dr. Robert Bowman, the former head of space weapons research, ISSS presents the real facts about SDI and why the space defense system not only wouldn't work but how it would fuel the arms race and actually bring us closer to war.

Dennis L. Haughton, M.D.

Interhelp
P.O. Box 331
Northamton, MA 01061
(413) 586-6311

An international network based on the conviction that peace, justice, and a healthy planet demands more than politics alone. They assist people in understanding their feelings about the dangerous trends that imperil our planet and show how we can begin acting effectively to create a peaceful future. My most recent trip to the Soviet Union was organized by Interhelp, and during it we explored some powerful and yet simple techniques for bridging understanding between Soviets and Americans that could be used with any group.

Mobilization for Survival.
853 Broadway, #418
New York, NY 10003

A nationwide organization of grassroots groups working for disarmament, non-intervention, safe energy, and human needs. It publishes *The Mobilizer.*

The Peace Child Foundation
3977 Chain Bridge Road
Suite 204
Fairfax, VA 22030

Arranges for diverse versions of the play *Peace Child* to be performed by teenagers around the world.

Physicians for Social Responsibility
1601 Connecticut Avenue, N.W.
Suite 800
Washington, D.C. 20009

The U.S. affiliate of IPPNW (International Physicians for the Prevention of Nuclear War).There is no effective treatment for nuclear war, only prevention. PSR has provided valuable education and increased public awareness of the consequences of nuclear war and shown that there would be no medical treatment available for the vast majority of survivors of global nuclear war.

SANE/FREEZE
711 G Street, S.E.
Washington, D.C. 20003

It is a nationwide grassroots disarmament organization. It publishes SANE World/FREZE Focus.

Search for Common Ground.
2005 Massachusetts Avenue, N.W.
Lower Level
Washington, D.C. 20036

An organization that acts as a catalyst to help other institutions take on dramatic projects with U.S. - Soviet participation such as producing TV documentaries on such topics as: "U.S. and Soviet Citizens Talk Via Satellite About Three Mile Island and Chernobyl".

Dennis L. Haughton, M.D.

Patricia Sun
P.O. Box 7065
Berkley, CA 94707-0065

Founder of ICU (The Institute of Communication for Understanding), travels the U.S. giving workshops on personal empowerment, relationships, going beyond fear to freedom, communication and love, self-actualization, and moving beyond duality. She has been a powerful catalyst in my own growth and I would recommend contacting her Berkley office to see when she may be in your area.

Union of Concerned Scientists
26 Church Street
Cambridge, MA 02238
(617) 547-5552

Its quarterly publication *Nucleus* is a nationally respected advocate of arms control, nuclear power safety, and national energy policy. UCS is an organization of scientists and other citizens concerned about the impact of advanced technology on society and makes available to its members a wealth of revealing information relevant to the issues discussed in this book.

BIBLIOGRAPHY

The following is a list of the principal sources used in writing this book as well as interesting reading. This list is by no means complete, but is presented with the intention of giving you a start for your own exploration.

Bach, Richard. ***Illusions: The Adventures of a Reluctant Messiah.*** **New York: Dell Publishing Co., 1977.** This is a delightful story about Richard's encounter with a modern day master and how we materialize into our lives whatever we hold in our thoughts. Uplifting and inspiring, this book teaches us to see beyond the illusions of the world and believe in ourselves again.

Bach, Richard. ***The Bridge Across Forever: A Lovestory.*** **New York: William Morrow and Co., Inc., 1984.** My Favorite by Bach. Anyone interested in relationships, soulmates, out-of-body experiences, and immortality will enjoy reading this book about Richard's personal journey of transformation that goes a step beyond *Illusions*.

Bloomfield and Cain and Jaffe. ***T.M.: Transcendental Meditation. Discovering Inner Energy and***

Overcoming Stress. **New York: Delacorte, 1975.** This is a good overview of meditation from the theory of how it works to the scientific research that documents its personal and sociological benefits.

Bloomfield and Kory. ***The Holistic Way to Health and Happiness: A New Approach to Complete Lifetime Wellness*. New York: Simon and Schuster, 1978.** This is an excellent book that shows how health is a positive state of being gained through self-actualization and not just the absence of disease.

Bucke, Maurice, M.D. ***Cosmic Consciousness: A Study in the Evolution of the Human Mind*. New Hyde Park, N.Y.: University Books, Inc., 1966.** This is a classic work about the next level of consciousness that awaits humanity illustrated by writings of the enlightened masters of the past who exhibited cosmic consciousness in their lives.

Capra, Fritjof. ***The Tao of Physics*. Boston: New Science Library, 1985.** Written by a physicist, this book explores the striking parallels between the underlying concepts of modern physics and Eastern mysticism. The understanding of reality gained from each viewpoint is amazingly similar.

Carlson, Don and Comstock, Craig, eds. ***Citizen Summitry*. New York: St. Martin's Press, 1986.** Contains essays focusing on solutions. There are five pieces on getting to know the Soviets, six by Soviets on space-bridges, six on transforming our consciousness, and six on envisioning the future and the use of conflict management.

"Chernobyl - One Year After." ***National Geographic*, Vol. 171, No. 5 (May 1987): pp. 632-659.** This is a well done and concise account of the Chernobyl disaster and its aftermath, including some excellent photographs and illustrations.

English, Robert D. and Halperin, Jonathan J. *The Other Side: How Soviets and Americans Perceive Each Other*. New Brunswick, USA: Transaction Books, 1987. This illuminating book gives valuable insights into how Soviets and Americans see one another based on distortions and prejudices gained from books, movies, and the media during the last seventy years of ideological warfare.

Foell, Earl and Henneman, Richard, eds. *How Peace Came to the World*. Cambridge: MIT Press, 1986.Contains a few dozen of the best essays that were sent in to the *Christian Science Monitor* in response to a contest called "Peace 2010". In it writers describe how peace was achieved from the perspective of someone living in the year 2010.

Fromm, Erich. *The Art of Loving*. New York: Bantam Books, 1963. This is a classic book on just what love is all about that takes it from a vaguely defined romantic feeling to a well defined art that can be developed to its full potential. As the most basic force underlying all healthy human relationships - from self love to universal love - it is important that we understand how to cultivate it.

Gorbachev, Mikhail S. *Perestroika: Our Hopes for Our Country and the World*. New York: Harper & Row, 1987. Divided into two parts, Gorbachev's latest book is devoted to plans and hopes for changes within the Soviet Union as well as views, expectations and policies toward the rest of the world.

Gorbachev, Mikhail S. *A Time for Peace*. New York: Richardson and Steirman, 1985. This is a collection of public statements made by General Secretary Gorbachev between March and October 1985 that gives the reader valuable insight into this world leader's thinking about events in the Soviet Union and the world in general.

Gorbachev, Mikhail S. ***Toward a Better World.*** **New York: Richardson and Steirman, 1987.** This contains Mr. Gorbachev's views as documented in public statements from February to December 1986. The first chapter is a personal message to American readers that is the healthiest appraisal of the world situation that I have seen from any political leader. Mr. Gorbachev and I share many of the same ideas of peace that are discussed in this book.

Hawkes, Nigel, et al. ***Chernobyl, The End of the Nuclear Dream.*** **New York: Vintage Books, 1987.** This is the most complete description of the events connected with Chernobyl I have found. Fascinating to read for anyoone interested in Chernobyl and the nuclear situation in general, it was the main source of facts for Chapter Four.

Kennan, George F. ***The Nuclear Delusion.*** **New York: Pantheon Books, 1976.** Written by an outspoken former ambassador to the Soviet Union, this book traces the historical, psychological, and political forces that have shaped the present arms race. In light of this understanding, the current justifications for needing huge nuclear arsenals and defense establishments to protect us from one another lose most of their validity.

Keyes, Ken, Jr. ***Handbook to Higher Consciousness.*** **Coos Bay, Ore.: Living Love Publications, 1975.** For anyone new to the ideas of higher consciousness this book is a good beginning, with many workable techniques for transforming yourself and your world into a more joyful existence.

Keyes, Ken, Jr. ***The Hundredth Monkey.*** **Coos Bay, Ore.: Vision Books, 1981.** At $2.00 per copy and uncopyrighted so you can duplicate it for your friends, this book shows - through the principle of the hundredth monkey (described in Chapter Eight of this book) - how we have the power to bring about a quantum leap in human consciousness and put an end to the nuclear arms race and bring about a better world.

MacLaine, Shirley. *Out on a Limb*. New York: Bantam Books, 1983, and *Dancing in the Light*. New York: Bantam Books, 1985. In her writing, Shirley shares the intimate story of her own personal transformation during her intense search for self and the reason for life. This soul searching spiritual quest finally leads her to her Higher Self and a new enlightened understanding of life, one that can inspire all of us to seek for ourselves.

***Nuclear Times*. [Washington, D.C.]** A magazine published six times a year by Nuclear Times, Inc. It contains a variety of articles on U.S. - Soviet relations and the peace movement.

Orme-Johnson, David, Ph.D. and Farrow, John, Ph.D., eds. *Scientific Research on the Transcendental Meditation Program: Collected Papers, Vol. 1*. Weggis, Switzerland: Maharishi European **Research University Press, 1977.** The best single publication of TM research, through 1977. Volumes Two and Three are completed, but not yet published.

***Peacemakers*. [Tallahassee, Fla.]** A magazine published monthly by Loiry Publishing Group (3380 Capital Circle N.E., Tallahassee, Fla., 32308). It contains hard news, information about upcoming actions and meetings, a review of recent events, profiles of available speakers, policy debates, profiles of people and organizations making a difference, reviews of the latest books and films, and more.

Pearson, Durk and Shaw, Sandy. *Life Extension*. New York: Warner Books, 1982. If you have a serious interest in longevity and living an active and healthy life to age one hundred twenty, then this book is for you. Durk and Sandy present very convincing evidence that by applying knowledge gained through longevity research to our daily living, we can add many healthy years to our lives.

Polyson, James, Ph.D. *Hard Rain: Nuclear War in Quotes*. Tallahassee, Fla.: Loiry Publishing House, 1988. This useful reference volume is a collection of approximately three thousand quotes from well-known individuals, historical and contemporary, expressing facts, opinions, hopes, fears, and ways of dealing with the threat of nuclear holocaust.

Sivard, Ruth Leger. *World Military and Social Expenditures 1986*. Washington, D.C.: World Priorities, 1986. Revised yearly, this is a storehouse of facts documenting how the global arms race consumes scarce and vital resources that are unavailable for pressing social needs. The facts speak for themselves and show how we often give armaments higher priorities than people. This was the source for most of the facts, statistics, and numbers concerning the arms race and social expenditures throughout this book.

***Surviving Together*. [Washington, D.C.]** A magazine published three times a year by the Institute for Soviet American Relations, (1608 New Hampshire Avenue, N.W., Washington, D.C., 20009). It contains articles about what's happening in both countries with U.S. - Soviet relations along with hundreds of short descriptions of cooperative ventures between the two countries that have been recently announced.

Time Life Books. *The Soviet Union*. Alexandria, Va.: Time Life Books, 1985. The Time Life staff does an excellent job in presenting an objective and unbiased overview of the Soviet Union. A good place to start for anyone wishing to become more informed about the USSR.

Willis, David K. *Klass: How Russians Really Live*. New York: St. Martin's Press, 1985. Behind the vision of an ideal society often portrayed in Soviet propaganda, lies the real working of a social system that has a long way to go to reach those goals. This book gives an excellent view into day to day Soviet life as it really occurs.